Next-Generation Technologies in Dairy Processing and Production

Poliana Mendes de Souza
University Mohammed VI Polytechnic, Morocco

IGI Global
Scientific Publishing
Publishing Tomorrow's Research Today

Vice President of Editorial Melissa Wagner
Director of Acquisitions Mikaela Felty
Director of Book Development Jocelynn Hessler
Production Manager Mike Brehm
Cover Design Jose Rosado

Published in the United States of America by
IGI Global Scientific Publishing
701 East Chocolate Avenue
Hershey, PA, 17033, USA
Tel: 717-533-8845
Fax: 717-533-7115
Website: https://www.igi-global.com E-mail: cust@igi-global.com

Library of Congress Cataloging-in-Publication Data

Names: Souza, Poliana Mendes de, 1984- editor
Title: Next-generation technologies in dairy processing and production / edited by Poliana Mendes de Souza.
Description: Hershey, PA : IGI Global Scientific Publishing, 2026. | Includes bibliographical references and index. | Summary: "The primary objective of the book is to provide a comprehensive overview of the latest technological advancements that are transforming dairy processing methods. This book seeks to bridge the gap between research and practical application, offering insights into how emerging technologies can enhance efficiency, safety, quality, and sustainability in the dairy industry"-- Provided by publisher.
Identifiers: LCCN 2025001368 (print) | LCCN 2025001369 (ebook) | ISBN 9798369399491 hardcover | ISBN 9798369399507 paperback | ISBN 9798369399514 ebook
Subjects: LCSH: Dairy processing--Technological innovations | Dairying | Agricultural innovations
Classification: LCC SF250.5 .N49 2025 (print) | LCC SF250.5 (ebook) | DDC 637.0285--dc23/eng/20250312
LC record available at https://lccn.loc.gov/2025001368
LC ebook record available at https://lccn.loc.gov/2025001369
British Cataloguing in Publication Data
A Cataloguing in Publication record for this book is available from the British Library.

Table of Contents

Detailed Table of Contents

The dairy business is not an exception to the ways artificial intelligence (AI) is changing other industries. Traditional methods in dairy farming are being revolutionized by the use of AI technology, which also improve sustainability, efficiency, and animal health. AI-driven technologies like computer vision, machine learning, and internet of things sensors are being utilized more and more for automation, precision farming, and data-driven decision-making. Animal health monitoring is a crucial use of AI in dairy farming. This chapter mainly concentrates on the benefits of using AI in dairy industries and also showcases the methods on how artificial intelligence can be implemented in the diary industries. The chapter also focuses on how AI is utilized in the dairy processing industry to improve quality control, optimize production procedures, and forecast demand. Artificial intelligence (AI)-driven robots help with cleaning, packing, and milking, which lowers labor costs and increases productivity.

Monitoring milk quality is essential for global food security due to its nutritional importance and high consumption. Milk remains vulnerable to adulteration and

chain technologies. These technologies services not only provide services to the dairy farms and dairy industries but also provide service to end customers with quality product delivery on time. Therefore, integration of these technologies is inevitable. This chapter provides the details of dairy farms, manufacturing dairy products, and IoT integration in dairy farms and in dairy industries.

The integration of internet of things (IoT) technologies into dairy processing facilities has the potential to revolutionize the industry by improving operational efficiency, product quality, and safety. This chapter explores the current state of IoT adoption in the dairy sector, focusing on the key technologies, benefits, challenges, and future trends. IoT sensors and actuators enable real-time monitoring and control of critical process parameters, whereas data analytics and machine learning algorithms facilitate predictive maintenance, quality control, and process optimization. Future trends and innovations in this field include the development of advanced sensors, improved data handling capabilities, and the integration of artificial intelligence for predictive analytics and decision support. To fully harness the potential of IoT in the dairy industry, strategic planning, collaboration among stakeholders, and continuous investment in research and development are essential.

The dairy industry is vital in the global agricultural economy, providing essential nutrition and significant economic contributions. However, it faces persistent challenges such as inefficiencies, waste, and environmental unsustainability. Traditional dairy processing systems, characterised by high resource consumption and limited technological integration, need help to meet modern requirements for efficiency and sustainability. The internet of things (IoT) has become a solution that offers real-time monitoring, predictive maintenance, process optimization, and traceability of supply chains. This chapter addresses IoT-driven automation in dairy processing, emphasizing the reduction of waste, increased productivity, and quality consistency. It discusses challenges in adopting IoT, including initial high costs, interfacing with legacy systems, data security issues, and strategies for overcoming these barriers. Through IoT-enabled systems, the dairy industry will be able to move toward a sustainable and connected future, meeting global demand with minimal environmental footprint.

This chapter examines the transformative potential of blockchain technology in the dairy sector, with a focus on camel milk as an emerging market. It explores how blockchain enhances traceability, quality assurance, and transparency while addressing challenges like inefficiencies, food safety, and inequities faced by small-scale producers. Tools such as IoT integration and smart contracts are highlighted for their role in streamlining operations and ensuring compliance with safety and sustainability standards. The chapter also discusses blockchain's ability to reduce intermediaries, promote fair trade, build consumer trust, and expand market access for niche products like camel milk. Challenges to adoption, including technological barriers, costs, and regulatory concerns, are explored alongside future prospects, such as innovations in scalability, emerging market expansion, and collaborative efforts between governments and industries to create a more equitable and sustainable dairy value chain.

The Moroccan bovine dairy industry is modernizing through digital technologies to enhance management and productivity. This study evaluates digitalization's current state, obstacles, and development opportunities, including the reliability of a 3D camera for automatic body condition scoring of dairy cows. Analysis of farms (15), technology suppliers (4), milk collection centers (9), and processing plants (4) reveals diverse adoption levels: all farms use surveillance cameras, 8 use automatic fans and misters, 5 have automatic weighing systems, 7 monitor mastitis with conductivity meters, and 4 use pedometers. Software like ISALAIT and Dairy Comp 30 support herd management. Digital technology suppliers offer innovative solutions and technical support, while milk collection centers and processing plants focus on traceability and automation. Financial and technical support and ongoing training are essential to accelerate digital transformation in the sector.

Preface

INTRODUCTION

The dairy industry is experiencing a pivotal transformation fueled by an urgent need for innovation, efficiency, and sustainability. As Editor of *Next-Generation Technologies in Dairy Processing and Production*, I am pleased to present this timely and comprehensive reference volume that brings together a range of insights into the cutting-edge technologies redefining dairy operations today.

The global dairy sector, long reliant on traditional methods, is undergoing a fundamental shift. Increasing consumer demand for safer, higher-quality, and more diverse dairy products—coupled with rising environmental and resource constraints—has catalyzed the adoption of advanced technologies. From intelligent sensors and automation to artificial intelligence, non-thermal processing, and blockchain integration, the tools of Industry 4.0 are no longer futuristic concepts; they are active agents of change reshaping the dairy landscape.

This volume aims to bridge the often-wide gap between research innovation and practical implementation in the field. Each chapter explores critical areas where next-generation technologies are making tangible impacts—from quality monitoring and processing optimization to smart packaging and waste reduction. A particular emphasis is placed on digital integration, highlighting how AI and IoT can revolutionize production systems by improving precision, reducing inefficiencies, and facilitating sustainable practices.

Our contributors—comprising leading researchers, practitioners, engineers, and academics—offer diverse perspectives and case-based insights that highlight both the challenges and opportunities facing the industry. Whether examining the role of extracellular polysaccharides in product innovation, or the digitization of national dairy sectors such as Morocco's, this book provides a panoramic yet detailed view of where the dairy industry is headed.

Next-Generation Technologies in Dairy Processing and Production is intended as a resource for a wide spectrum of readers. Industry professionals will discover strategies to enhance operational efficiency and product innovation. Academicians and researchers will find a wealth of material for further study and application. Engineers and technology developers will benefit from in-depth discussions on equipment design and process optimization. Sustainability advocates, policymakers, and consultants will find actionable insights on how technology can support eco-conscious production and regulatory compliance.

In assembling this volume, it has been my goal to provide not only a state-of-the-art reference but also a catalyst for further dialogue and innovation. I am confident that this book will serve as a valuable guide and inspiration for those dedicated to shaping a smarter, safer, and more sustainable dairy future.

ORGANIZATION OF THE BOOK

Chapter 1: Next-Generation Technologies in Dairy Processing and Production: Utilization of Artificial Intelligence in Dairy Operations

The dairy business is not an exception to the ways artificial intelligence (AI) is changing other industries. Traditional methods in dairy farming are being revolutionized by the use of AI technology, which also improve sustainability, efficiency, and animal health. AI-driven technologies like computer vision, machine learning, and Internet of Things sensors are being utilized more and more for automation, precision farming, and data-driven decision-making. Animal health monitoring is a crucial use of AI in dairy farming. This chapter mainly concentrates on benefits of using ai in dairy industries and also showcases the methods on how artificial intelligence can be implemented in the diary industries. The chapter also focuses on how AI is utilized in the dairy processing industry to improve quality control, optimize production procedures, and forecast demand. Artificial intelligence (AI)-driven robots help with cleaning, packing, and milking, which lowers labor costs and increases productivity.

Chapter 2: Modern Approaches to the Use of Sensors in Milk Quality Monitoring

Monitoring milk quality is essential for global food security due to its nutritional importance and high consumption. Milk remains vulnerable to adulteration and chemical changes despite technological advances in dairy production. Fraudulent practices such as dilution and addition of adulterants pose health risks and undermine trust. Traditional detection methods are fast and simple, but lack the sensitivity required for complex scenarios. This chapter explores modern approaches to milk quality monitoring, focusing on spectroscopy, mass spectrometry, colorimetry, chromatography, biosensors, and electrokinetic separation techniques. These advanced tools offer high sensitivity, real-time analysis, and portability, providing more reliable and efficient solutions. However, challenges such as high upfront costs and the need for specialized training limit widespread adoption. These technologies not only address current challenges but also prepare the dairy industry for sustainable reliability and operational efficiency in the future.

Chapter 3: Role of Extra Cellular Polysaccharides in Dairy Products

This chapter provides an extensive exploration of EPS, highlighting the types, functions, and applications in different dairy matrices. EPS in dairy products is mainly produced by lactic acid bacteria. It is essential to note that EPS can be linked to the texture, viscosity, and gelation properties of dairy products. These characteristics are decisive for thus organoleptic properties and consumers' desirability for food products, such as, e.g., yogurt, cheese, and ice cream. In addition to technological aspects, EPS is associated with numerous health benefits, such as probiotic effects, immune modulation, and potential prebiotic action. However, this chapter also focuses on the technological issues of both processing and utilizing EPS for dairy products. These elements should be further researched because they can help optimize use and facilitate the process of production of new, high quality food products that meet consumers' and regulatory agencies' requirements

Chapter 4: Integration of IoT in Dairy Processing Facilities

Internet of Things technology is found in almost of the all real life applications The applications include smart home, smart cities, smart grid etc. Similarly, a kind of scenario can be seen in the dairy sector also. Agriculture and dairy farms are the backbone of many countries, and their maintenance becomes laborious and expensive. To meet the afore mentioned criteria, one needs to depend on the latest technologies such as Internet of Things, Machine learning, artificial Intelligence, block chain and supply chain technologies. These technologies services not only provide services to the dairy farms and dairy industries but also provide service to end customers with quality product delivery on time. Therefore, Integration of these technologies is inevitable. This chapter provides the details of dairy farms, manufacturing dairy products and IoT integration in dairy farms and in dairy industries.

Chapter 5: Integration of IoT in Dairy Facilities

This chapter explores the current state of IoT adoption in the dairy sector, focusing on the key technologies, benefits, challenges, and future trends. IoT sensors and actuators enable real-time monitoring and control of critical process parameters, whereas data analytics and machine learning algorithms facilitate predictive maintenance, quality control, and process optimization. Future trends and innovations in this field include the development of advanced sensors, improved data handling capabilities, and the integration of artificial intelligence for predictive analytics and decision support. To fully harness the potential of IoT in the dairy industry, strategic planning, collaboration among stakeholders, and continuous investment in research and development are essential.

Chapter 6: IoT-Driven Automation in Dairy Processing for Waste Reduction and Productivity Enhancement

The dairy industry is vital in the global agricultural economy, providing essential nutrition and significant economic contributions. However, it faces persistent challenges such as inefficiencies, waste, and environmental unsustainability. Traditional dairy processing systems, characterized by high resource consumption and limited technological integration, need help to meet modern requirements for efficiency and sustainability. The Internet of Things (IoT) has become a solution that offers real-time monitoring, predictive maintenance, process optimization, and traceability of supply chains. This chapter addresses IoT-driven automation in dairy processing, emphasizing the reduction of waste, increased productivity, and quality consistency. It discusses challenges in adopting IoT, including initial high costs, interfacing with legacy systems, data security issues, and strategies for overcoming these barriers. Through IoT-enabled systems, the dairy industry will be able to move toward a sustainable and connected future, meeting global demand with minimal environmental footprint.

Chapter 7: Integration of Blockchain in Dairy Sector Value Chain

This chapter examines the transformative potential of blockchain technology in the dairy sector, with a focus on camel milk as an emerging market. It explores how blockchain enhances traceability, quality assurance, and transparency while addressing challenges like inefficiencies, food safety, and inequities faced by small-scale producers. Tools such as IoT integration and smart contracts are highlighted for their role in streamlining operations and ensuring compliance with safety and sustainability standards. The chapter also discusses blockchain's ability to reduce intermediaries, promote fair trade, build consumer trust, and expand market access for niche products like camel milk. Challenges to adoption, including technological barriers, costs, and regulatory concerns, are explored alongside future prospects, such as innovations in scalability, emerging market expansion, and collaborative efforts between governments and industries to create a more equitable and sustainable dairy value chain.

Chapter 8: The Digitization of the Moroccan Bovine Dairy Sector: Current Situation, Challenges, and Prospects

The Moroccan bovine dairy industry is modernizing through digital technologies to enhance management and productivity. This study evaluates digitalization's current state, obstacles, and development opportunities, including the reliability of a 3D camera for automatic body condition scoring of dairy cows. Analysis of farms (15), technology suppliers (4), milk collection centers (9), and processing plants (4) reveals diverse adoption levels: all farms use surveillance cameras; 8 use automatic fans and misters; 5 have automatic weighing systems; 7 monitor mastitis with conductivity meters, and 4 use pedometers. Software like ISALAIT and Dairy Comp 30 supports herd management. Digital technology suppliers offer innovative solutions and technical support, while milk collection centers and processing plants focus on traceability and automation. Financial and technical support and ongoing training are essential to accelerate digital transformation in the sector.

CONCLUSION

As we navigate an era marked by rapid technological evolution and growing environmental awareness, the dairy sector stands at the forefront of a vital transformation. The chapters in this volume collectively underscore the remarkable potential of next-generation technologies to address longstanding industry challenges while opening new avenues for innovation, quality enhancement, and sustainable development.

This book is not just a reflection of current advancements—it is a forward-looking resource intended to spark further research, collaboration, and implementation across academia, industry, and policy-making. My hope is that *Next-Generation Technologies in Dairy Processing and Production* will serve as a foundational text for professionals and scholars alike, inspiring meaningful change in how we process, monitor, and deliver dairy products to a global population.

Let this compilation be a call to action—for embracing technology not as a replacement for tradition, but as an enabler of a more resilient, efficient, and responsible dairy industry.

Poliana Mendes de Souza
University Mohammed VI Polytechnic, Morocco

Chapter 1

Next-Generation Technologies in Dairy Processing and Production:

Utilization of Artificial Intelligence in Dairy Operations

A. Mariamadonna
https://orcid.org/0009-0009-1561-9133
PSG College of Arts and Science, India

R. Saranya
PSG College of Arts and Science, India

ABSTRACT

The dairy business is not an exception to the ways artificial intelligence (AI) is changing other industries. Traditional methods in dairy farming are being revolutionized by the use of AI technology, which also improve sustainability, efficiency, and animal health. AI-driven technologies like computer vision, machine learning, and internet of things sensors are being utilized more and more for automation, precision farming, and data-driven decision-making. Animal health monitoring is a crucial use of AI in dairy farming. This chapter mainly concentrates on the benefits of using AI in dairy industries and also showcases the methods on how artificial intelligence can be implemented in the diary industries. The chapter also focuses on how AI is utilized in the dairy processing industry to improve quality control, optimize production procedures, and forecast demand. Artificial intelligence (AI)-driven robots help with cleaning, packing, and milking, which lowers labor costs

DOI: 10.4018/979-8-3693-9949-1.ch001

and increases productivity.

INTRODUCTION

The dairy industry is an important part of the global agricultural system, contributing to better nutrition, food security, and economic development. It was expected (according to FAO 2023) that, worldwide, milk production would reach 930 million metric tones by 2023, due to rapidly developing cities, population growth, and increased global demand for dairy products. Despite regional advances, many challenges still confront the dairy industry, including: inefficiencies, food waste, and sustainable environmental practices. As demand for dairy products continues to increase, continuing to develop answers for these challenges is necessary and urgent. Therefore, it is clear that the need for efficient and sustainable dairy production systems will increase (Kumar & Joshi, 2021). According to literature, it is estimated that approximately twenty percent of waste associated with dairy products (both milk and dairy derivatives) equates to millions of metric tons each year (Smith & Wang, 2020).

In addition to financial losses, wages may cause environmental issues such as rising resource consumption and greenhouse gas emissions. A real agricultural and environmental impact is being felt; to produce one litre of milk requires four to seven litres of water. Those major processes are also energy intensive as pasteurization, homogenization and packaging have a large carbon footprint (Libelium, n.d.; Gupta & Roy, 2023). And we must keep in mind that the dairy industry can be voluntary subjected to more scrutiny pressure from regulatory bodies to become a cleaner, more efficient, and less wasteful one.

IoT can resolve these problems in a way that they only way we can dream about. The IoT connects physical non-electronic devices, sensors, and data analytics to create an intelligent, connected system that is capable of monitoring performance, assessing decisions, and optimizing processes in real-time. From using intelligent supply chain systems in product distribution and waste management to gathering raw milk and quality assurance, IoT could transform the dairy processing value chain. For example, with the help of IoT sensors in milk collecting systems, sensors can constantly monitor variables such as temperature, pH, and total bacteria counts, and ensure that only top quality milk enters the processing chain (Lee & Lee, 2015). Processing facilities can dynamically modify production parameters thanks to IoT-based automation, which reduces material waste, increases energy efficiency, and guarantees improved product consistency (Fitcows, n.d.).

Besides, the IoT promotes predictive maintenance, which prolongs asset life and prevents downtime due to breakdowns. IoT-based smart waste management systems track waste generation in real-time in dairy plants and assist in identifying areas of waste or inefficiency. These technologies lessen the environmental impact of dairy processing by recycling and reusing materials, further optimizing water and energy use (Arya et al., 2023a; Arya et al., 2023b). The application of useful knowledge from IoT and AI, along with advanced data analytics enhance sustainability and decision making, and productivity (Libelium, n.d.; Mhapsekar et al., 2024).

The global dairy market will be almost $500 billion by 2022 and will have a compound annual growth rate (CAGR) of 5% until 2030.To match the demand for this sector while responding to environmental and economic pressure, adopting IoT-driven automation becomes imperative, not an option. Improving efficiency in operations, cutting resource waste, and ensuring quality products through IoT can change the fabric of traditional dairy processing and position the industry towards a sustainable future (Lee & Lee, 2015; Arya et al., 2023a).

In terms of waste reduction and efficiency improvement, this chapter examines the revolutionary potential of IoT-driven automation in dairy production. It addresses IoT applications at critical points in the dairy production process, presents successful implementation case studies from the real world, and looks at the difficulties in implementing IoT. The chapter concludes by discussing potential avenues for future study and innovation and offering a road map for academic, business, and policymakers to use IoT to create a more sustainable and effective dairy sector.

THE DAIRY INDUSTRY: AN OVERVIEW

Milk is one of the largest components of the economy for food worldwide. It provides livelihoods for millions of farmers and their families and nutrition for billions of human beings. The movement of movements makes it feasible to efficiently and effectively move milk, cheese, yogurt, butter and powdered milk from whisking or harvesting to processing to distribution. With its protein, its calcium, and its vitamins, the milk industry supports global nutrition and supports the wellness of the dietary requirements of millions of people worldwide. Milk accounts for nearly 10% of global agricultural production. According to the FAO (FAO, 2023), milk is the third most valuable agricultural commodity behind rice and wheat. And in 2023, the world consumed more than 930 million metric tons of milk. Given population growth, urbanization, and increases in per capita levels of consumption, demand for milk is expected to sharply increase.

ICAR estimates that India, alongside the United States, China, and Brazil, is the world's largest producer of milk, with the current production representing more than 23% of total global production (Zarin et al., 2023).Inefficient resource use, environmental impacts, and sustainability requirements—all of which are sustained by rising global demand—present a challenge to the industry despite its strategic significance (Ahmad et al., 2019).

The dairy industry offers massive economic contributions to agriculture; it is the foundational tool for rural development and jobs. Smallholder dairy producers significantly contribute to milk production volumes, especially in developing regions. On the other hand, milk production in industrial countries involves large-scale farms and modern processing plants; there is a domestic demand for and export of such products. Gross Domestic Products in many countries benefit hugely from the dairy industry while anchoring food security and economic stability.

Dairy Industry Structure

The dairy industry is a complex value chain involving several stages and stakeholders. It can be broadly categorised into two primary segments: Production, processing and Distribution.

Primary Production

This stage includes the production of raw milk at the farm level. Key activities include:

- Dairy farming: There are milk-producing animals such as goats, sheep, cows, and buffaloes. The authors Galukande et al. (2013) stated that animals health care, feeding and breeding practices are main factors that determine high yields and quality of milk.
- Milk Collection: Bulk tanks store milk mainly at rural collection centers or cooperative societies. Quality tests include percentage fat content, pH levels, and bacterial count at collection time.
- Transportation: Milk is transported to processing units under controlled conditions to prevent spoilage, usually in refrigerated tankers.

Primary production differs significantly from region to region. Smallholder farmers dominate developing countries, whereas large-scale, automated dairy farms are more common in developed countries. Modern farms use automated milking systems, GPS cattle tracking, and feed optimization for improved efficiency.

Processing and Distribution

This stage processes raw milk into consumable products and distributes them to the final consumers. It includes:

Processing:

- Pasteurization: Milk is heated to kill harmful microorganisms and increase shelf life.
- Homogenization: Breaking down fat globules makes the texture uniform and increases stability.
- Fermentation: It is about the process of making yoghurt, cheese, and other cultured dairy products.
- Drying and Condensing: This That process can also produce powders and processed milk products (condensed milk etc.) that are designed to be stored for a long time and/or exported.
- Packaging: Advanced packaging procedures and advanced food science approaches are used to protect products and increase shelf life and potentially increase value.

Distribution:

- Distribution to consumers via retail chains, supermarkets, and direct-to-consumer channels.
- Cold chain logistics for transporting and storing products (Aung & Chang, 2014).

Key Players in the Dairy Industry

The dairy value chain is made up of several stakeholders with a broad variety of interests:

- Farmers: Producing milk is the first stage of quality control.
- Cooperatives: In terms of gathering milk, guaranteeing its quality, and serving as a market for small-scale farmers, they are also the most important.
- Processing Companies: Utilizing cutting-edge technologies, turn raw milk into a variety of dairy products.

- o Retailers and Distributors: Assure consumers that processed goods are delivered while upholding quality requirements.
- o Consumers: Encourage the market for a variety of sustainable, high-quality dairy products.

CHALLENGES IN TRADITIONAL DAIRY PROCESSING

Traditional processing practices have historically been the foundation of the dairy industry, but traditional processing is no longer meeting modern measures of effectiveness, sustainability and quality. Traditional systems face numerous challenges as global demand for dairy products rise, leading to lower productivity, increased waste and biodiversity impacts. The key challenges of traditional dairy processing are:

Resource Intensity

To continue giving milk, dairy farming requires significant energy, feed, land, and water. Water is needed for cleaning, milk processing, and hydration, so recycling water and using it wisely are paramount to waste reduction. Well-managed land is necessary for both animal grazing and the feed crops, and sustainable land management can minimize soil loss. The dairy sector also relies on high energy consuming transportation, refrigeration, and milking equipment, but any savings and environmental cost reduction can be reduced using renewable energy sources - like solar and biogas. While alternatives like precision feeding and alternate protein sources are increasing efficiencies, more resources are still needed for feed production.

- o High Water Consumption: Water is essential to the cleansing, chilling, and processing processes of traditional dairy production. Four to seven liters of water are thought to be needed to digest one liter of milk. Freshwater resources are strained by this excessive consumption, especially in areas where water is scarce (FAO, 2023; Ahmad et al., 2019).
- o Energy-Intensive Processes: Pasteurisation, homogenisation, and refrigeration consume substantial energy. Traditional systems' dependence on non-renewable energy sources further increases their environmental footprint (Gupta & Roy, 2023).

Waste Generation

Manure, wastewater, and leftover feedstuff are the major sources of waste generation in dairy production, and they present a real threat to the environment. As organic matter with high nutrient value, manure management can become a pollution burden that can diminish soil and water quality when not properly handled. Both organic and inorganic matter in wastewater must be treated before it can be discharged. To add to this issue, wasted feed as a result of spoilage or uneaten feed also creates waste and wastes feed resources unnecessarily. Waste management processes such as composting, biogas and precision feeding practices are used on-farm to manage waste and limit the amount of excess. Sustainable practices like reuse of water or conversion of manure to fertilizers or energy, are examples of practices that have the potential to limit environmental impacts and enhance farm efficiency. Solid Waste: Dairy processing creates large quantities of solid waste. This includes a large variety of waste products. Spoiled milk, waste products, scraps, and packaging can all be considered waste.Shamsuddoha and Nasir (2024) explain the economic implications of disposing of these products that are not being managed and becoming pollution.

- Liquid Waste: High concentrations of organic debris, lipids, and nutrients are found in cleaning, washing, and processing wastewater. These can harm the environment by contributing to eutrophication and water contamination if released untreated (Ahmad et al., 2019).

Quality and Safety Issues

Quality and safety concerns are problems that have evidently impacted product integrity in relation to consumer health and the reputation of the dairy sector. There are connection issues that can arise from a multitude of issues; contamination hazards, improper handling, poor sanitary environments, hazards presented by harmful and/or pathogenic bacteria such as Salmonella, E. coli and many others. In order to ensure milk safety, pasteurization, in-store storage and sanitary regulations at both the farm level and processing facilities must be adhered to if milk safety is to be successfully supported; there are also pesticide and antibiotic residues that can compromise milk quality, both of which are regulated under food safety regulations. Unethical acts such as milk adulteration can also lead to harmful situations, and therefore, transparency and traceability matter. Safety and efficacy monitoring, utilizing such means as blockchain technology and automated quality control systems and testing can create some support to quality monitoring, safety monitoring, and compliance needs.

- o Inconsistent Product Quality: Inconsistent product quality results from traditional systems' inability to continuously monitor and regulate vital parameters including temperature, pH, and microbial activity (Fitcows, n.d.).
- o Food Safety Risks: The risk of contamination, spoiling, and noncompliance with food safety regulations is increased by inadequate monitoring. Consumer trust may be impacted, leading to product recalls and monetary losses (Dabbene et al., 2014; Aung & Chang, 2014).

Inefficient Supply Chains

Inefficient supply chains in the dairy sector contribute to delays, waste, extra costs, and reduced profitability. Spoilage and quality loss of milk products are commonly caused by a lack of cold chain management and inadequate transport infrastructure, and by poor real-time tracking. Timely and efficient delivery of dairy products to consumers may also be impacted by out-of-date logistics systems, as well as dispersed delivery and distribution networks. Ineffective inventory management compounds the waste when there are under-supplied situations or over-supplied situations due to ineffective forecasting and market demand signals. To address these challenges, modern dairy farms and companies are implementing data-sourcing competitive optimization, block chain technology for trust, and automated logistics systems as means to increase efficiency within their supply chain.

- o Milk Spoilage: Raw milk often spoils because of poor transport from farms to processing plants with delays or poor transportation assets (cold storage) and poor supply chain cohesion or disconnects in the supply chain.
- o Traceability Challenges: Full end-to-end product traceability is also a challenge in traditional systems and therefore it is difficult to identify cause(s) of issues in the supply chain or confirm the validity of products.

Equipment Downtime

Inefficient supply chains can cause huge waste and cost to the dairy industry, such as spoilage of milk, increased operating costs, and delays. Disconnected distribution channels, outdated logistics technology, and a lack of effective cold chain control affect quality of the product and profitability. Farmers and distributors will disrupt the market if they experience shortages or overproduction because of inaccurate inventory predictions. Additionally, supply chain issues due to lack of tracking systems and congestion can limit consumer access to fresh dairy products. Dairy Farms are turning to blockchain technology for transparency, AI forecasting tools, and automated logistics to mitigate these issues, increase productivity, decrease waste, and maintain timeliness.

- o Unplanned Maintenance: Unplanned Maintenance: Typically, this type of equipment does not allow for predictive maintenance capabilities, leading to a greater number of breakdowns and unplanned downtime, which disrupts production schedules and adds more operational expenses (Mhapsekar et al., 2024).
- o Aging Infrastructure: Many dairy processing plants rely on outdated machinery that is less efficient and more prone to failures, further hindering productivity (Nestlé Sustainability Reports, 2021).

Environmental Impact

The primary environmental impacts of dairy production should be land degradation, consumption of water, and greenhouse gas emissions. Methane is a very potent greenhouse gas and contributor to climate change, so it is very important to mitigate and maintain emissions as would any greenhouse gas. Dairy farmers should attempt to reduce their water use by implementing water conserving practices, even though they do not only require water for the animals, but even more cleaning the animals and facilities, and also water to produce animal feed. This includes, both recycling wastewater and utilizing an efficient irrigation system for crops. The employment of land for feed production and grazing will also need to employ stewardship to maintain the land from degradation and deforestation, as this has relation to both climate change and sustainability.

- o Greenhouse Gas Emissions: Through energy consumption, wastewater management and transportation, the dairy processing sector is a key source of greenhouse gas (GHG) emissions (Libelium, n.d.). Commonly used methods of processing dairy products do not typically have controls to reduce or offset GHG emissions.
- o Excessive By-Products: Without advanced utilisation techniques, by-products such as whey are often discarded or underutilised, contributing to resource wastage (Ahmad et al., 2019).

Labor Dependency

In dairy production, labor dependence is an integral part of maintaining output, livestock management, and operational efficiency. For an acceptable level of efficiency in performing everyday tasks on conventional farms, skilled labor is also important because traditional farms in particular are labour-intensive. Labour is needed to milk cows, prepare feeds, provide health care, and for the other everyday jobs. Nonetheless, the company experienced some difficulties a couple years ago because of labour shortages, increased costs, and the fact that caring for cows is a therapeutic career for dairy workers.As a result, farms are now increasingly using automated systems and new technologies such as AI herd monitoring, automated feeding stations, and robotic milking in an effort to improve productivity and reduce their reliance on labour.

- Manual Operations: Traditional dairy processing heavily depends on manual labour for monitoring, controlling, and adjusting processes. This limits scalability and increases the risk of human error (Mahalik & Nambiar, 2010).
- Workforce Limitations: Labor shortages, particularly in rural areas, can disrupt operations, and reliance on unskilled labour can compromise quality and efficiency (Galukande et al., 2013).

Limited Data Utilization

Efficiency, productivity, and sustainability are all hampered by the underutilization of data in dairy production. Many farms continue to use conventional techniques without fully utilizing data-driven insights, even in the face of agricultural technological improvements. Ineffective herd management, milk production, feed optimization, and illness control are caused by inadequate data collection and processing. If farmers don't track and monitor their cows properly then there will be problems planning that result in wasted time, inefficient use of resources, and will cost more money and produce less. IoT sensors, AI-supported analytics, and automatic measurement mean farmers can monitor herd health, monitor logistics, and maximize profit. By knowing, and using data, a dairy farm can increase efficiency, improve sustainability, and decrease waste.

- Lack of Real-Time Monitoring: Insufficient or Absence of Real-Time Monitoring. Conventional systems include manual periodic inspections, which often delays responses and may introduce inaccuracies causing lost time and interactions with equipment and product quality.
- Inefficient Data Use: The absence of integrated data systems prevents operators from gaining insights into performance trends, resource usage, and areas for improvement.

Regulatory and Market Pressures

Various market and regulatory factors influence the agricultural outcomes, profit, sustainability and even the future of dairy businesses. Government regulations around environmental compliance, animal welfare regulations, food safety regulations, as well as documentation and record keeping imposes requirements on farms that often increase production costs. In addition, dairy producers are required to innovate their operations and must also consider the financial viability of changing consumer demands, changing milk prices, and the threat of competition from plant-based alternatives. As consumers are demanding food that is more transparent, ethically sourced, and sustainably produced; farms are undergoing the implementation of new technology and sustainable practices in agriculture. To remain viable and competitive in an ever changing and dynamic food system, dairy farms must align compliance, profits, and changes in consumer demand. Increasing Compliance Requirements: As global food safety and environmental regulations become more stringent, traditional systems struggle to adapt due to limited monitoring and reporting capabilities (Dabbene et al., 2014).

- Growing Consumer Expectations: Modern consumers demand high-quality, sustainably produced dairy products. Traditional methods often fail to meet these expectations, resulting in competitive disadvantages (Lee & Lee, 2015).

THE ROLE OF TECHNOLOGY IN TRANSFORMING THE DAIRY INDUSTRY

Technological advancements in dairy are changing the scenario for these challenges. The Internet of Things (IoT) encourages smart and sustainable dairy processing. IoT allows real-time monitoring of production parameters, prediction of equipment maintenance, optimization of resources, which directly reduces waste and enhances productivity (Mhapsekar et al., 2024; Libelium, n.d.). Automated systems of dairy processing ensure consistency in quality.

Also, IoT applications impact the primary production by using a smart farming (for example, automatic milking robots, GPS enabled cattle tracking, and feed optimization) which are changing dairy farming companies. These new technological advances improve animal health and welfare, milk output and overall efficiency, providing an almost uninterrupted supply of high-quality milk to the milk processors.

The Internet of Things (IoT) is changing the dairy processing, with real-time monitoring, predictive maintenance, and process optimization capabilities in the dairy processing industry. These IoT capabilities improve greenhouse gas emissions, grass losses, operational inefficiencies as well as providing a quality product (Nagraj et al., 2022; Almaiah et al., 2023). Figure 1. represents an IoT-controlled automatic dairy processing system.

Figure 1. IoT-driven automatic dairy processing

Real-Time Monitoring of Milk Quality

IoT technologies ensure that quality monitoring is consistent across all stages of dairy processing. Real-time sensors track critical parameters, for example: temperature, pH, and microbial counts, to assure regulatory compliance and product safety (FAO, 2023; Fitcows, n.d.). The Food and Agriculture Organization estimate that average global milk spoilage rates are approximately 20% (FAO, 2023). The integration of IoT has been shown through literature that they can reduce spoilage previously estimated at 15-20%. This is largely due to improved storage and transportation systems. Table 1 represents the key parameters monitored in IoT-driven dairy processing system.IoT sensors enable real-time monitoring of critical milk quality factors, thus ensuring safety and consistency across all processing steps-

- o **Temperature Control**: Milk must always be kept at the proper temperatures to avoid spoilage. IoT-based temperature sensors provide real-time data, meaning corrective measures can be taken immediately if the temperature deviates. Implementing IoT-based temperature monitoring systems has reduced spoilage rates by 20% (Fitcows, n.d.; Libelium, n.d.).
- o **pH Levels**: Checking pH levels reveals contamination and ensures product uniformity. IoT sensors allow continuous pH measuring, thus allowing a timely response to any adverse changes (Smith & Wang, 2020).
- o **Fat and Protein Content**: Checking milk's nutritional ingredients is important for quality monitoring. IoT devices can measure fat and protein content in real-time and ensure adherence to product specifications (Gupta & Roy, 2023).

Table 1. Key parameters monitored in IoT-driven dairy processing

Parameter	**Optimal Range**	**IoT Application**	**Improvement in Quality (%)**
Temperature	2°C – 4°C	Real-time cooling system monitoring	18%
pH Level	6.5 – 6.8	Automated pH sensors	15%
Microbial Count	< 50,000 CFU/mL	Microbial detection systems	20%

Predictive Maintenance of Equipment

IoT facilitates predictive maintenance by monitoring equipment health indicators and minimizing downtime and maintenance costs (Amul Dairy Cooperative, 2022; Mhapsekar et al., 2024). Figure 2. represents the reduction in equipment downtime with IoT-based predictive maintenance.

- o Vibration Analysis: Sensors detect abnormal vibrations in machinery, indicating potential failures. Early detection allows for timely maintenance, preventing costly breakdowns.
- o Temperature Monitoring: Increases in temperature in equipment parts can indicate wear or malfunction. IoT devices track such changes, allowing preemptive intervention (Fitcows, n.d.; Kumar & Joshi, 2021).
- o Energy Consumption: Energy usage patterns indicate inefficiencies. IoT systems alert operators to anomalies, and energy-saving interventions can be undertaken (Nestlé Sustainability Reports, 2021).

Figure 2. Reduction in equipment downtime with IoT-based predictive maintenance

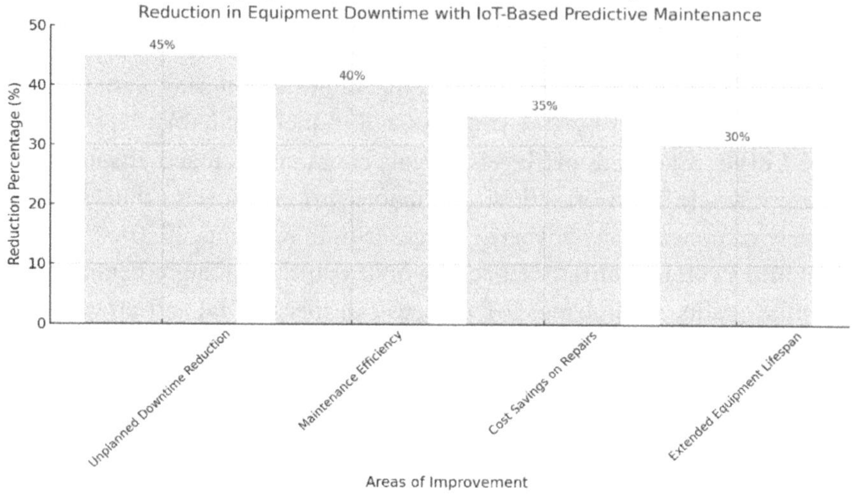

Process Optimization

IoT offers the possibility of optimizing dairy processing operations through real-time data and control.

- o Control of Pasteurization: Temperature control is critical to ensure product safety in pasteurization. IoT systems enable precise temperature control, making it more efficient and consistent (Gupta & Roy, 2023).
- o Homogenization Efficiency: In homogenization, the pressure and flow rate should be maintained constant to ensure uniform texture in the product. IoT devices help adjust for optimal conditions (Mahalik & Nambiar, 2010).
- o Fermentation Monitoring: In yogurt and cheese manufacturing, IoT sensors monitor acidity and temperature to achieve desired product characteristics (Ahmad et al., 2019).

Smart Feeding Technology

Smart feeding technology is revolutionizing the dairy industry in herd health improvement, waste reduction, and nutrition management, through data-driven automation. The adoption of traditional feeding practices contains inefficiencies like overfeeding and underfeeding, which affect digestion, milk supply, and farm costs. Smart feeding systems tackle these problems by measuring a cow's dietary need according to her age, weight, milk yield, and overall health. With the help of automated feeders IoT sensors, these systems calculate and dispense portions according to each cow's need with the help of Artificial Intelligence powered analytics. Intelligent systems enable automated modifications to feeding schedules to avoid nutritional deficits and guarantee maximum output by adjusting to real-time feeding pattern shifts. These systems offer farmers integrated software to monitor consumption patterns, anomalies, and feeding schedules, further increasing productivity and sustainability. Smart feeding systems enable real-time monitoring of livestock feed intake, nutritional content, and wastage via Internet of Things enabled Smart devices. This helps them provide optimal nutrition to the animals with minimal waste of feeds. Sensors like GPS trackers and automated feeders yield data which the smart feeding technology uses for adjustment as shown in Figure 3.

- Improved Feed Efficiency: Reduces overfeeding and underfeeding, optimizing the feed-to-milk conversion ratio.
- Cost Savings: Reduces feed wastage and lowers overall feeding costs by up to 10–15% (Gupta & Roy, 2023; Galukande et al., 2013).
- Environmental Benefits: Minimizes methane emissions from livestock by optimizing feed composition (Mahalik & Nambiar, 2010).

Figure 3. Smart feeding technology

Waste Reduction and Resource Management

IoT in dairy processing significantly reduces waste and conserves resources. Table 2. shows the impact of IoT on resource utilization.

- Water Consumption Monitoring: IoT sensors measure water consumption in real-time, thus identifying where it could be saved. This has reduced water consumption by 15% in some dairy plants (Fitcows, n.d.; FAO, 2023).
- Energy Management: IoT systems analyze energy consumption patterns to optimize usage, save costs, and reduce negative environmental impacts.
- By-Product Utilization: Process monitoring can identify valuable by-products for efficient use or recycling (Shamsuddoha & Nasir, 2024).

Table 2. Impact of IoT on resource utilizatin

Resource	Traditional Usage	IoT-Enabled Usage	Reduction (%)
Water	7 litres /litre of milk	5.5 litres/litre	20%
Energy	15 kWh/ton	12.5 kWh/ton	15%
Raw Material	1.2 kg/litre of product	1.1 kg/litre	8%
Feed Wastage	25%	10%	15%

Supply Chain Traceability

IoT Traceability across the dairy supply chain ensures high quality and safety from farm to consumer (Aung & Chang, 2014; Dabbene et al., 2014). Figure 4. represents the improvement in supply chain transparency with IoT implementation.

- RFID Tagging: RFID tags attached to the milk can be tracked throughout the whole supply chain, thus enabling transparency and accountability (Gupta & Roy, 2023; IoT ONE, n.d.).
- Blockchain Integration: By integrating the IoT with the blockchain network, data is secured for an immutable record of the product's movement (Mhapsekar et al., 2024).
- Quality Assurance: Real-time information about storage and transit conditions ensures that product quality is maintained, and spoilage is not at risk (Libelium, n.d.).

Figure 4. Improvement in supply chain transparency with IoT implementation

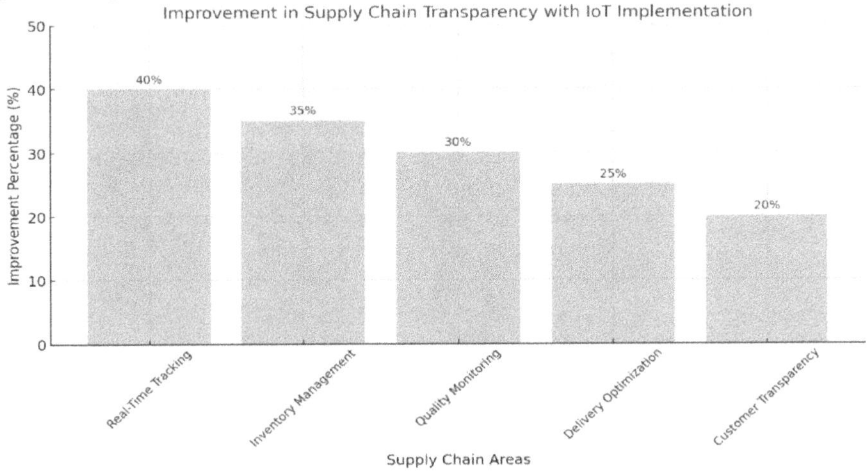

PROPOSED METHODOLOGY

The process begins with initiating IoT-enabled monitoring and automation systems in the dairy plant. This will provide seamless integration of sensors, devices, and analytics platforms. Figure 5. depicts the flowchart of the process of IoT-enabled monitoring and automation systems in the dairy plant.

Figure 5. Flowchart of the process of IoT-enabled monitoring and automation systems in the dairy plant

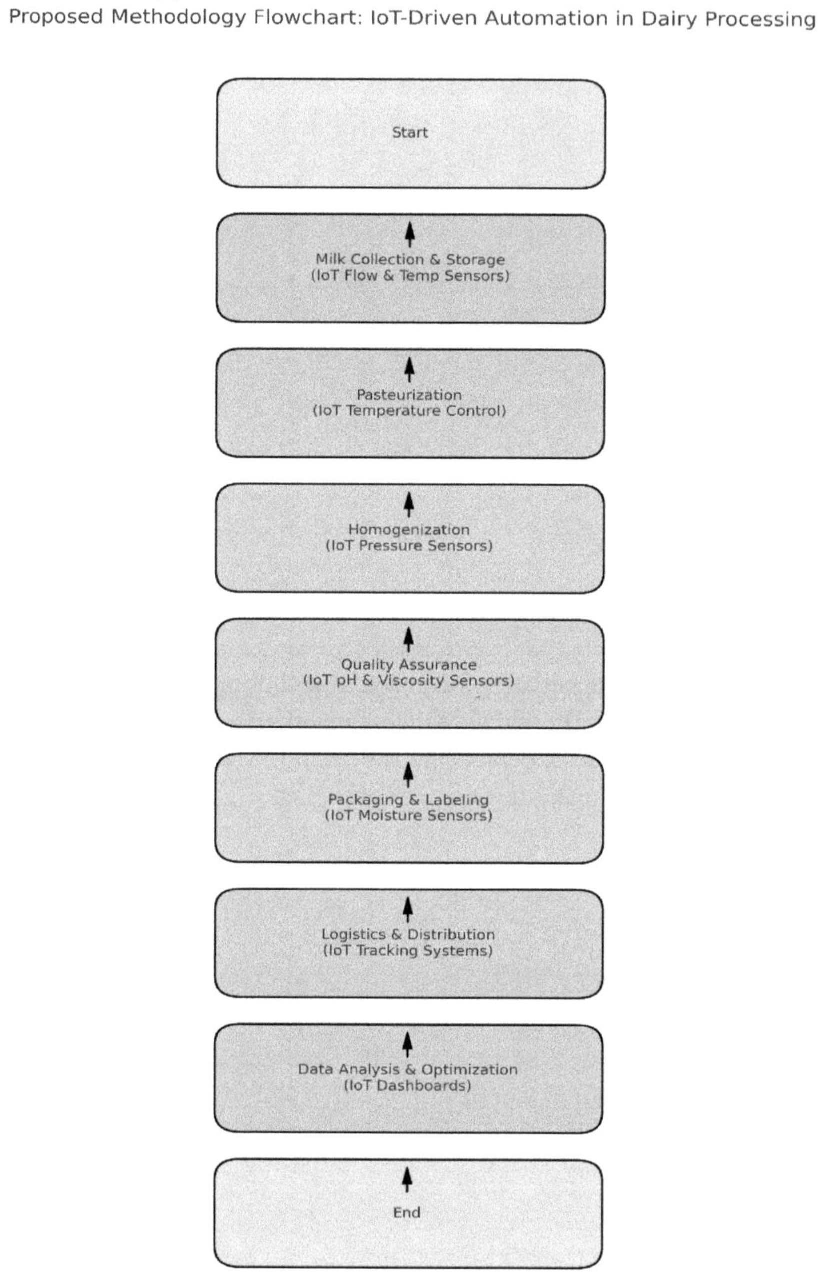

The flowchart outlines the steps in IoT-driven automation in dairy processing while focusing on waste reduction and productivity enhancement:

Collection & Storage

- **IoT Flow and Temperature Sensors**: IoT flow sensors track the incoming volume during milk collection to avoid spillage or overflows. Temperature sensors ensure milk stays fresh under optimal conditions (Arya et al., 2023a; Kumar & Joshi, 2021).
- **Real-Time Monitoring**: Immediate Monitoring: Sensor information can be sent to a centralised location, so that any irregularities can be addressed immediately through corrective action being taken to a deviation like a temperature change (Fitcows, n.d.; FAO, 2023).

Pasteurization

When we consider existing dairy products, pasteurization makes a responsible claim about the safety of the milk for human use as it destroys harmful germs and maintains the nutritional content of milk. Milk is heated to a temperature for a period of time to destroy germs such as Salmonella, E. coli and Listeria.After heated it is cooled. Pasteurization keeps the milk flavour and quality and ensures a longer shelf life. There are a number of methods used including high-temperature short-time (HTST) and low-temperature long-time (LTLT) pasteurization, which vary based on finding a balance between efficient and safe.Dairy farmers help the public by lowering the danger of foodborne illnesses and guaranteeing that consumers receive fresh, uncontaminated dairy products through the use of pasteurization.

- **IoT Temperature Control**: Pasteurization is carried out on the milk to kill harmful bacteria yet preserve nutrients. IoT sensors precisely monitor and control this process, thereby maintaining quality while adhering to food safety requirements (Smith & Wang, 2020).
- **Automation**: IoT-enabled systems can automate heating cycles, allowing optimal energy use and wastage reduction (Nestlé Sustainability Reports, 2021).

Homogenization

Homogenization is a very important step in dairy processing. It causes a continuous uniformity in milk by keeping the cream from separating. Raw milk not only has fat in it but the fat globules will float. The process of homogenization homogenizes the fat globules into little parts dispersed uniform throughout the milk by applying high-pressure. The distinctions produced from homogenization are extensive and invisible. For example, appearance, better taste, shelf-life and a smoother mouthfeel. Homogenization stabilizes milk, improves digestibility of milk while retaining its full nutritional value, is one more crucial step of modern dairy production and is an allocated process that, when separated from pasteurization and performed independently, allows people to rely on a uniform product to receive high standard milk.

- **IoT Pressure Sensors**: Homogenization breaks fat molecules into a uniform texture. Pressure sensors help the process to run at optimum levels and avoid under or over-processing (Mahalik & Nambiar, 2010).
- **Efficiency**: Real-time adjustments do not waste resources and ensure product consistency (Amul Dairy Cooperative, 2022).

Quality Assurance

Quality assurance is an important aspect in the production of dairy products to provide assurance regarding the quality, safety and consistency of dairy products. Quality assurance refers to various processes or steps, such as the hygienic collection of milk, proper control of storage conditions, and compliance with regulatory requirements. Farmers use advanced testing methods to measure the composition of milk, detect contaminants, and quantify bacterial levels. Regulatory agencies also have stringent regulations related to the process of pasteurization, homogenization, and packaging to maintain food safety. Technology is essential; blockchain traceability and automated monitoring systems increase consumer trust and transparency. Dairy farms can produce premium milk and dairy products while preserving the public's health and the industry's reputation by putting strict quality assurance procedures in place.

- **IoT pH and Viscosity Sensors**: The quality of milk or processed dairy products is assessed at this stage. Sensors with IoT capabilities monitor the pH levels, viscosity, etc., to ensure these products meet the quality criteria (Libelium, n.d.; Gupta & Roy, 2023).
- **Automated Feedback Loops**: Any deviation sensed by the sensors gives an alert or adjusts the product to avert human error.

Packaging & Labeling

The dairy industry uses packaging and labeling to protect quality, health, and communicate with consumers. Packaging for dairy products incorporates materials like plastic, glass, and cartons to contain freshness, minimize exposure to contamination, and create advantages for shelf life. New sustainable packaging technologies - aseptic packaging, compostable and biodegradable materials - will further sustainability and maintain product integrity.

- **IoT Moisture Sensors**: Moisture sensors ensure that packaging materials are proper and the product has not been contaminated. The labelling process is also automated, ensuring accuracy and traceability.
- **Waste Reduction**: Automated packaging wastes less material and increases productivity.

Logistics & Distribution

Dairy product packaging and labeling is an important aspect of the dairy industry as it protects the quality of the product, maintains food safety, plus provides necessary information about the product to the user. Environmental protective factors such as plastics, glass and cartons can preserve shelf life, reduce the possibility of cross contamination, and preserved freshness.Some examples of innovations that move the focus to sustainability while not compromising the parcel and user quality include biodegradable plastics along with aseptic packaging.

- **IoT Tracking Systems**: Upon product readiness, IoT-enabled tracking monitors temperature and humidity during transportation to guarantee its integrity (IoT ONE, n.d.; Aung & Chang, 2014).
- **Supply Chain Optimization**: IoT systems provide real-time location tracking and updates, thus improving supply chain transparency and efficiency (Dabbene et al., 2014).

Data Analysis & Optimization

Data analysis and optimization are initiating a revolution in the dairy industry by increasing productivity, sustainability and profitability. Using advanced technologies -- artificial intelligence (AI), machine learning, and Internet of Things (IoT) sensors -- farmers are collecting large quantities of data related to milk yields and animal health, feed consumption and management, and environmental conditions.

- **IoT Dashboards**: All collected data is aggregated and analysed on IoT dashboards. Insights from this data drive process optimisation, such as improving resource allocation, reducing energy consumption, and identifying bottlenecks (Gupta & Roy, 2023; Mhapsekar et al., 2024).
- **Predictive Analytics**: The system predicts maintenance needs and process improvements, thus ensuring continuous productivity.

The process results in an efficient and minimized waste operation with IoT technologies that consistently met quality and reduced operational costs, and optimized on-time delivery. This process outlines how IoT-enabled automation is contributing to improved dairy processing with real-time monitoring, automation, and data-enhanced optimization. It ensures sustainability and reduces waste, thus fulfilling modern industry needs.

CASE STUDIES OF IOT IMPLEMENTATION IN DAIRY PROCESSING

The case study discusses how IoT innovations have transformed the operational setup of dairy industries worldwide through downtime, resource optimisation, milk production, and general efficiency.

Amul, India, implemented IoT-based predictive maintenance systems and automated quality monitoring to enhance equipment efficiency and reduce downtimes. The outcomes were remarkable, with a 25% reduction in downtime, 30% savings in maintenance costs, and a 20% improvement in milk quality consistency (Amul Dairy Cooperative, 2022). Nestlé, Europe focused on sustainability by deploying IoT-enabled sensors for water recycling, energy optimization, and by-product utilization. This strategy resulted in considerable resource optimization, reduced water usage by 20%, increased energy consumption by 15%, and successfully repurposed whey into high-value protein supplements (Nestlé Sustainability Reports, 2021). At Voshazhnikovo Farm, Russia, IoT sensors were used to monitor environmental conditions and cow behaviour. Integrated with machine learning, the system gave an 18% rise in milk production and €340,000 in cost savings over 180 days on feed, indicating that technology in dairy farming drastically changes productivity (Libelium, n.d.). Stellapps Technologies, India, introduced IoT applications to improve productivity and profitability in dairy farming. Real-time monitoring and managing dairy operations led to measurable improvements in operational efficiency and financial outcomes (IoT ONE, n.d.). The Edge-Centric Industrial IoT Solution demonstrated the power high technologies such as Edge-AI and smart sensors can bring to optimise dairy processing activities. It featured real-time monitoring and control capabilities, dramatically improving productivity and efficiency (Mhapsekar et al., 2024). Table 3 presents case study examples of operational applications in dairy processing using IoT-driven automation and their objectives, implementation, and outcomes.

Table 3. Case studies of operational applications in dairy processing using IoT-driven automation

Case Study	Objective	Implementation	Outcome	Reference
Amul, India	Reduce equipment downtime and improve efficiency	IoT-based predictive maintenance systems and automated quality monitoring.	- Reduced equipment downtime by 25%. - Maintenance costs lowered by 30%. - Milk quality consistency improved by 20%.	Amul Dairy Cooperative, 2022
Nestlé, Europe	Optimize resource usage and enhance sustainability	IoT-enabled sensors for water recycling, energy optimization, and by-product utilization.	- Water usage reduced by 20%. - Energy consumption decreased by 15%. - Whey repurposed into high-value protein supplements.	Nestlé Sustainability Reports, 2021
Voshazhnikovo Farm, Russia	Increase milk production and optimize feeding efficiency	Deployment of IoT sensors to monitor environmental conditions and cow behavior, combined with machine learning algorithms to analyze data	- 18% increase in milk production - €340,000 savings in feed costs over 180 days	Libelium, n.d.
Stellapps Technologies, India	Enhance productivity and profitability in dairy farming	Implementation of IoT-based milk production applications to monitor and manage dairy operations	- Improved productivity and profitability in dairy farming	IoT ONE, n.d.
Edge-Centric Industrial IoT Solution	Optimize dairy processing operations	Integration of IoT, Edge-AI, and smart sensors for real-time monitoring and control	- Enhanced productivity and efficiency in dairy processing	Mhapsekar et al., 2024

CHALLENGES IN IOT IMPLEMENTATION IN DAIRY PROCESSING

Automating the dairy processing and manufacturing sectors using IoT has many potential benefits. However, IoT implementation has associated technical, operational, and cost issues that may inhibit adoption and scale. This section outlines aspects of IoT technology implementation's challenges specific to the dairy sector.

High Initial Costs

Capital investment: IoT systems incur substantial capital investments in sensors, devices, software, and networks. Most of the dairy industry is represented by small and medium enterprises (SMEs), and can afford the capital investments potential/ kumar & Joshi, 2021). IoT system-building in dairy plants increases operational costs in the initial years by 15-20% (FAO,2023).

Data Security and Privacy

As dairy operations will be using digital technology for automated operations, supply chain optimization, and herd management, the security and privacy of their data has become vital. Dairy operations accumulate extensive data of a sensitive nature, for example financial transactions, animal health data and records of milk production, using cloud-based platforms, IoT sensors, and AI analytics. Protecting such information from cyber-attacks, unauthorized access, and breaches of information, is necessary for preserving farm integrity, and compliance with industry regulations. With adequate encryption methods, secure access, and regular cybersecurity assessments, elements of their digital assets can be protected.

- Cybersecurity Risks: IoT devices are prone to cyberattacks, including data breaches, ransomware, and unauthorized access. All these risks can compromise the business and product data (Mhapsekar et al., 2024).
- Regulatory Compliance: Compliance with data protection regulations, such as GDPR and industry-specific standards, makes it more complex to implement the IoT (Libelium, n.d.).

Integration with Legacy Systems

- Compatibility Issues: Most dairy processing facilities use aged equipment and infrastructure, and integrating modern IoT systems with them is challenging (Amul Dairy Cooperative, 2022).

- o Operational Disruptions: Integrating IoT functionality into existing systems creates short-term process disruptions (Nestlé Sustainability Reports, 2021).

Technical Competence Gaps

- o Skill Gaps: Operating and maintaining IoT systems demands advanced technical skills in sensor deployment, data analytics, and cybersecurity. Dairy operators lack this kind of skill.
- o Training Costs: Up-skilling the employees to handle IoT systems increases the operational costs and time spent.

Connectivity Challenges

- o Rural Locations: Dairy farms and processing units are often located in remote areas with limited internet connectivity, which hinders real-time data transmission and system efficiency (FAO, 2023).
- o Latency Issues: Poor network performance can lead to delays in data collection and analysis, reducing the effectiveness of IoT systems (IoT ONE, n.d.).

Resistance to Change

- o Cultural Barriers: Many stakeholders in the dairy industry, especially in traditional setups, resist adopting new technologies because they do not trust or understand IoT(Lee & Lee, 2015).
- o Adoption Rates: The resistance to change slows the shift to IoT-enabled systems, especially in developing regions (Gupta & Roy, 2023).

Scalability Concerns

- o Infrastructure Limitations: Expanding IoT solutions across large-scale dairy operations can strain infrastructure and resources.
- o Cost vs. Benefit: The long-term benefits of IoT investments are difficult to prove in many cases, especially for small-scale operators.

Environmental and Power Constraints

- o Energy Requirements: IoT devices consume power continuously, which may be a limitation in energy-scarce or remote locations (Mahalik & Nambiar, 2010).

- o Environmental Durability: IoT sensors and devices must endure extreme temperatures and humidity levels typical in dairy processing environments (Dabbene et al., 2014).

Addressing the Challenges

1. Cost Reduction: Incentivizes governments or organizations regarding adopting IoT for the dairy industry or offering subsidies .
2. Improvement of Connectivity LPWAN or Satellite-Based IoT: Low-power wide area network technologies have improved connectivity in places with little coverage.
3. Training workforce industry players can create workforce development programs that are technically skilful.
4. Cyber Security To protect data, Strong encryption procedures should be enforced, and systems should be audited regularly (Mhapsekar et al., 2024).

CONCLUSION & FUTURE DIRECTIONS

The dairy area is at a crossroads and is still in a development phase, with challenges arising from increasing global demand for dairy products while figuring out how to address the urgent challenges presented by the inefficiencies, waste and environmental impacts associated with current processing systems. This chapter discussed the transformational nature of IoT-based automation systems in dairy processing operations, which improve productivity, respond efficiently to resource use, and limit waste. The world of IoT does not have comparable footprints and can optimize processes to primarily minimize the environmental footprint, while also allowing improvement of the product quality, while not reducing consumer expectations for local and sustainable food products, including real time monitoring of milk quality, guidance for predictive maintenance of equipment. The case studies from this chapter framed the real and tangible benefits of promoting IoT over the existing systems such as reduced processing duration with less downtime for equipment, improvements in resources utilization, and product quality measurements. However, major hurdles still remain to be overcome in order for full integration of IoT technologies to occur, including the challenge of funding (e.g. start-up costs) for implementation of these applications, challenges with technical integration, and site-specific data privacy issues; these barriers need to be addressed by the allocation of resources by policy-makers, industry stakeholders, and researcher.

Future directions for IoT-based automation in the dairy processing sector will see the integration of a variety of emerging technologies, including IoT (the Internet of Things), to improve efficiency, sustainability, and quality assurance. Automation systems can be combined with artificial intelligence (AI) and machine learning (ML) models to provide newer capabilities, including predictive maintenance on machinery, real-time assessment of production parameters, and improved volume and availability of supply chain logistics. Blockchain technology also has a potential application for traceability and transparency through the dairy supply chain and can help regain consumer trust and regulatory compliance. Digital twins (the virtual representation of real systems) can help dairy processing organizations conduct simulation-based testing to allow operators to assess and identify bottlenecks to improve process efficiencies without disrupting production. Further collaboration between IoT platforms and robotics in packaging and distribution can further streamline operations with reduced human intervention and consistency in product quality. Finally, these innovations will only scale if they are embedded with extreme cybersecurity measures to protect sensitive data and standardized IoT protocols in their adoption. These innovations have the promise to transform traditional dairy processing into a more connected, automated, and sustainable sector.

REFERENCES

Ahmad, T., Aadil, R. M., Ahmed, H., Rahman, U., Soares, B. C. V., Souza, S. L. Q., Pimentel, T. C., Scudino, H., Guimarães, J. T., Esmerino, E. A., Freitas, M. Q., Almada, R. B., Vendramel, S. M. R., Silva, M. C., & Cruz, A. G. (2019). Treatment and utilization of dairy industrial waste: A review. *Trends in Food Science & Technology*, *88*, 361–372. DOI: 10.1016/j.tifs.2019.04.003

Almaiah, M. A., Yelisetti, S., Arya, L., Christopher, N. K. B., Kaliappan, K., Vellaisamy, P., & Alkdour, T. (2023). A novel approach for improving the security of IoT–medical data systems using an enhanced dynamic Bayesian network. *Electronics (Basel)*, *12*(20), 1–15. DOI: 10.3390/electronics12204316

Amul Dairy Cooperative. (2022). *IoT implementation report: Efficiency and maintenance*. Internal Publication.

Arya, L., Sharma, Y. K., & Kumar, R. (2023). Towards a greener tomorrow: IoT-enabled smart environment monitoring systems. *2023 International Conference on Advances in Computation, Communication and Information Technology (ICAICCIT)*, 1112-1117. DOI: 10.1109/ICAICCIT60255.2023.10465894

Arya, L., Sharma, Y. K., Kumar, R., Padmanaban, H., Devi, S., & Tyagi, L. K. (2023). Maximizing IoT security: An examination of cryptographic algorithms. *2023 International Conference on Power, Energy, Environment & Intelligent Control (PEEIC)*, 1548-1552. DOI: 10.1109/PEEIC59336.2023.10451210

Aung, M. M., & Chang, Y. S. (2014). Traceability in a food supply chain: Safety and quality perspectives. *Food Control*, *39*, 172–184. DOI: 10.1016/j.foodcont.2013.11.007

Dabbene, F., Gay, P., & Tortia, C. (2014). Traceability issues in food supply chains: A review. *Biosystems Engineering*, *120*, 65–80. DOI: 10.1016/j.biosystemseng.2013.09.006

Fitcows. (n.d.). Milk monitoring system using IoT to check milk's quality. Retrieved from https://www.fitcows.com/blogs/milk-monitoring-system-using-iot/

Food and Agriculture Organization (FAO). (2023). Global milk production statistics. Retrieved from https://www.fao.org/dairy-production-products/production/milk-production/en

Galukande, E., Mulindwa, H., Wurzinger, M., Roschinsky, R., Mwai, A. O., & Sölkner, J. (2013). Cross-breeding cattle for milk production in the tropics: Achievements, challenges, and opportunities. *Animal Genetic Resources*, *52*, 111–125. DOI: 10.1017/S2078633612000471

Gupta, P., & Roy, S. (2023). Enhancing dairy productivity with IoT. *Advances in IoT Systems*, *5*(4), 299–315.

IoT ONE. (n.d.). IoT-based milk production application. Retrieved from https://www.iotone.com/case-study/iot-based-milk-production-application/c602Foo

Kumar, A., & Joshi, R. (2021). IoT in dairy: A sustainable approach. *International Journal of Dairy Technology*, *74*(3), 487–495.

Lee, I., & Lee, K. (2015). The Internet of Things (IoT): Applications, investments, and challenges for enterprises. *Business Horizons*, *58*(4), 431–440. DOI: 10.1016/j.bushor.2015.03.008

Libelium. (n.d.). How a dairy farm increased their milk production 18% with IoT and machine learning. Retrieved from https://www.libelium.com/libeliumworld/success-stories/how-a-dairy-farm-increased-their-milk-production-18-with-iot-and-machine-learning/

Mahalik, N. P., & Nambiar, A. N. (2010). Trends in food packaging and manufacturing systems and technology. *Trends in Food Science & Technology*, *21*(3), 117–128. DOI: 10.1016/j.tifs.2009.12.006

Mhapsekar, R. U., O'Shea, N., Davy, S., Kilbane, D., & Abraham, L. (2024). An edge-centric industrial IoT solution for smart dairy processing. *IEEE Internet of Things Magazine*, *7*(5), 80–87. DOI: 10.1109/IOTM.001.2400010

Nestlé Sustainability Reports. (2021). Water and energy optimization in dairy processing facilities. Retrieved from https://www.nestle.com

Shamsuddoha, M., & Nasir, T. (2024). Utilizing dairy waste processing for organic agricultural production: A sustainable approach to producing organic goods. *Processes (Basel, Switzerland)*, *12*(11), 2521. DOI: 10.3390/pr12112521

Smith, L., & Wang, Z. (2020). Waste reduction in dairy processing through IoT automation. *Journal of Sustainable Agriculture*, *12*(2), 200–214.

Zarin, I., Talukder, A. K., Hoque, M. N., Rahman, M., Haider, G., Rahman, A. N. M. A., & Das, Z. C. (2023). Current scenario on reproductive management and performances of dairy cows in Manikganj district. *Annals of Bangladesh Agriculture*, *27*(2), 129–144. DOI: 10.3329/aba.v27i2.72541

Chapter 2

Modern Approaches to the Use of Sensors in Milk Quality Monitoring

Flaviana Coelho Pacheco
https://orcid.org/0000-0002-1302-8059
Federal University of Viçosa, Brazil

Irene Andressa
https://orcid.org/0009-0005-5180-4695
Federal University of Viçosa, Brazil

Jeferson
https://orcid.org/0000-0001-6722-9051
Federal University of Viçosa, Brazil

Fabio Ribeiro dos Santos
https://orcid.org/0000-0001-8822-9421
Federal University of Viçosa, Brazil

Kely De Paula Correa
https://orcid.org/0000-0002-0595-7990
Cândido Tostes Dairy Institute, Agricultural Company of Minas Gerais, Brazil

Ana Flávia Coelho Pacheco
https://orcid.org/0000-0001-7220-1432
Cândido Tostes Dairy Institute, Agricultural Company of Minas Gerais, Brazil

Paulo C. Henrique Costa Paiva
https://orcid.org/0000-0002-5455-7790
Cândido Tostes Dairy Institute, Agricultural Company of Minas Gerais, Brazil

Bruno Ricardo de Castro Leite Júnior
Federal University of Viçosa, Brazil

ABSTRACT

Monitoring milk quality is essential for global food security due to its nutritional importance and high consumption. Milk remains vulnerable to adulteration and chemical changes despite technological advances in dairy production. Fraudulent practices such as dilution and addition of adulterants pose health risks and under-

DOI: 10.4018/979-8-3693-9949-1.ch002

mine trust. Traditional detection methods are fast and simple, but lack the sensitivity required for complex scenarios. This chapter explores modern approaches to milk quality monitoring, focusing on spectroscopy, mass spectrometry, colorimetry, chromatography, biosensors, and electrokinetic separation techniques. These advanced tools offer high sensitivity, real-time analysis, and portability, providing more reliable and efficient solutions. However, challenges such as high upfront costs and the need for specialized training limit widespread adoption. These technologies not only address current challenges but also prepare the dairy industry for sustainable reliability and operational efficiency in the future.

INTRODUCTION

Monitoring milk quality ensures global food security (Haldar et al., 2022). Recognized as a nutrient-dense and widely consumed food, milk contributes significantly to the human diet (Kumar et al., 2023). Technological advances in dairy farming have contributed considerably to production efficiency, incorporating modern practices such as advanced management techniques, optimized nutritional strategies, and effective disease control methods, resulting in increased productivity per animal (Patil et al., 2024). However, due to its highly perishable nature and vulnerability to microbiological contamination, the introduction of adulterants, and chemical changes along the supply chain make it essential to develop effective solutions to ensure their safety and quality (Ionescu et al., 2023).

Food fraud, defined as a practice that is economically motivated to obtain financial benefit, represents a growing concern (Moyer et al., 2017). In the case of milk, the high demand for this product and its derivatives encourages fraudulent practices (Lal et al., 2022). One of the most common practices is the dilution with water, followed by the addition of low-cost compounds with no nutritional value, such as sugar, salt, starch, and sodium carbonate, to mask the reduction in quality. In addition, adulteration occurs intentionally, unintentionally, or naturally (Brooks et al., 2021).

Unintentional tampering is commonly attributed to failures in handling or storage and a lack of adequate infrastructure to meet quality standards (Valand et al., 2020). Intentional adulteration, on the other hand, involves deliberately inserting harmful substances to increase the volume or artificially improve the product's appearance. On the other hand, natural adulteration stems from organic compounds in food, which can harm human health (Valand et al., 2020).

The consumption of adulterated milk poses serious health risks, including the development of serious illness, hospitalizations, and, in extreme cases, death (Azad & Ahmed, 2016). Kidney problems are particularly alarming among the consequences of consuming contaminated milk (Lal et al., 2022). These risks are amplified by the absence of adequate detection technologies and the lack of knowledge of good practices in the management and administration of drugs in the dairy industry (Robson et al., 2020).

Simple chemical processes based on color changes can be used to qualitatively detect adulterants in milk. Although these methods have advantages such as simplicity, speed, and ease of execution, they have significant limitations, including low accuracy and applicability restricted to specific concentration ranges, which reduces their effectiveness in more complex scenarios. Still, these techniques can be useful in situations that demand quick responses (Poghossian et al., 2019).

The growing demand for high-quality dairy products has highlighted the importance of modern sensors in ensuring food safety and efficiency in milk production. These advanced tools, including spectroscopy-based, colorimetric, biosensor, and electrochemical sensors, offer high sensitivity, compact design, and real-time functionality, making them ideal for detecting adulterants and contaminants. Despite challenges like high implementation costs, technical training needs, and maintenance demands, the long-term benefits—such as reduced losses, improved safety, and enhanced operational efficiency—justify their adoption. Additionally, integrating these sensors with big data and artificial intelligence opens new possibilities for predictive analytics and informed decision-making, solidifying their role as indispensable in the future of the dairy industry.

THE IMPORTANCE OF MILK QUALITY MONITORING

The quality of raw milk is decisive for all stages of dairy production, directly impacting food security and human health. In addition to reflecting the health, welfare, and nutritional status of the animals, its composition makes it essential to carry out safety and quality tests, which are essential both in dairy farming and in the processing industry (Poghossian et al., 2019; Tsenkova & Muncan, 2022).

Protein, lactose, and fat are the main constituents of milk, directly influencing its nutritional value, commercial, and processing efficiency for the production of dairy products (Tsenkova & Muncan, 2022). However, the quality of these constituents can be impaired by the absence of good management and manufacturing practices during milk obtainment, processing, storage, and transportation (Ionescu et al., 2023). Under these conditions, adulteration and fraud are common, such as the addition of water or chemical substances to mask quality deficiencies, compromising not only the physicochemical characteristics of the milk, but also food safety and consumer confidence (Sharma et al., 2017). However, it is also common for fraud to be carried out by some producers to obtain greater added value and profit (Mafra et al., 2022).

Milk is among the most frequently adulterated foods worldwide, mainly due to its high perishability, the imbalance between supply and demand in urban areas, and the lack of efficient methods for fraud detection (Tanzina & Shoeb, 2014; Ferreira et al., 2024). The neglect of these practices is compounded by the lack of strict enforcement by food safety authorities. Therefore, it becomes essential to implement more robust anti-fraud measures, especially during supplier audits, to ensure product integrity and protect the health of consumers (Ionescu et al., 2023).

Food analysis, from raw materials to the final product, is essential for food safety, serving industries, regulatory bodies, scientific institutions, and consumers (Galanakis, 2021). Food authentication is one of the main steps in this control, allowing the quality of the product to be assessed, the conformity of the information on the label to be verified, and to confirm aspects such as geographical origin, genetic composition, and production methods (Ferreira et al., 2024). In addition, authentication plays a crucial role in identifying fraud, including tampering, mislabeling, and false declarations of origin, which compromise public health and consumer trust (Danezis et al., 2016; Spink, 2019). Therefore, rigorous analysis is essential to ensure the integrity of food and protect all parties involved in the supply chain.

FRAUD AND ADULTERATION IN THE DAIRY SECTOR

Fraud and intentional adulteration of milk aims to mask its inferior quality or replace natural components for technical or commercial reasons, in addition to increasing its volume (Nascimento et al., 2016). Generally, this type of adulteration involves diluting the milk and/or adding cheap, low-quality products (Figure 1) (Afzal et al., 2011). This practice represents a widespread concern, as it can result in the loss of the nutritional value of milk and, at the same time, constitute a significant health risk, as the added substances may pose health risks, rendering the milk unfit for consumption (Sharma et al., 2017).

Figure 1. Most common forms of adulteration in milk

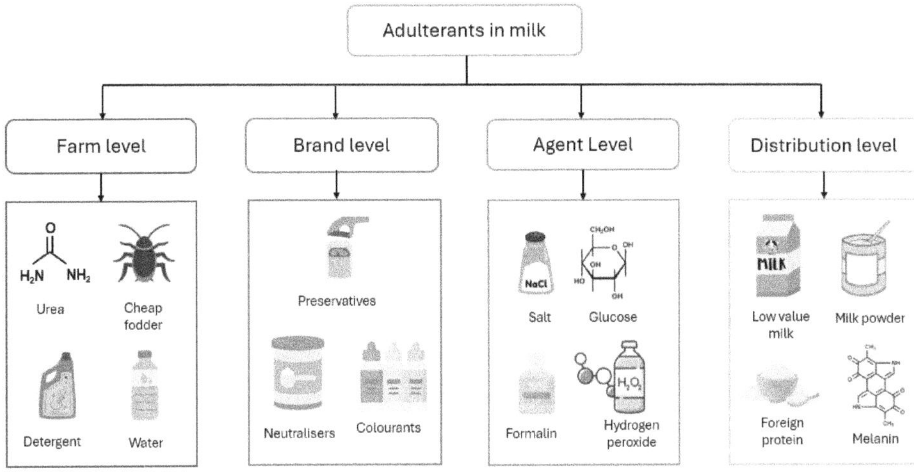

The most common and simple intentional milk fraud and adulteration consists of the addition of water, with the main objective of increasing volume, and can be easily detected using the cryoscopic freezing point method. However, when the water used is contaminated, significant health risks arise for consumers (He et al., 2019). On the other hand, the addition of nitrogenous compounds to increase the apparent protein content represents a more sophisticated and difficult to detect form. Examples of this practice include the incorporation of melamine, urea, and whey. Such adulteration is frequent, as traditional methods of protein quantification, such as Kjeldahl's, cannot distinguish between protein and non-protein nitrogen (Nascimento et al., 2016). In addition, these compounds have low cost and high nitrogen content, making fraud economically advantageous for fraudsters.

The use of chemicals is a common practice in milk adulteration. Substances such as hydrogen peroxide, formaldehyde, dichromate, boric acid, salicylic acid, and hypochlorite are often added to extend the shelf life of the product (Patil et al., 2024). Detergents are also employed to emulsify and dissolve water and oil, resulting in a foamy mixture that simulates desirable characteristics of milk (Yakubu et al., 2022). In addition, ammonium sulfate is used to maintain the density of diluted milk by manipulating the values obtained by the lactometer. Sodium carbonate[REMOVED HYPERLINK FIELD], sodium hydroxide and sodium bicarbonate are examples of neutralizers widely used to mask the high acidity of milk obtained under poor hygienic conditions. This high acidity comes from microbial growth during the obtainment, storage and transport of raw milk, especially in regions where milk is transported for a long period without adequate cooling facilities.

Glucose, or glucose syrup, is generally used to create a sweeter taste, as well as increase the solids content in milk (Stefas et al., 2021). Additionally, vegetable oils, such as soybean, sunflower, coconut, and peanut oils, animal oils, such as beef tallow and lard, and surfactants are used to artificially increase the fat content of milk (Qin et al., 2022). Detecting adulteration with vegetable oils is especially challenging due to the variation in their chemical composition. Adulteration with fats of animal origin presents additional difficulty, as the chemical composition of these fats is very similar to that of milk fat (Rebechi et al., 2016).

In addition to the deliberate adulteration of milk and its derivatives, falsification and incorrect labeling of the product can occur, as they can mislead the consumer by offering incomplete or misleading information about the product (Visciano & Schirone, 2021). These practices compromise not only the safety and quality of the item sold, but also consumer confidence in the brand or the market as a whole, and can lead to economic losses and damage to health. In addition, such acts constitute violations of current consumer protection and quality control legislation, and are subject to legal sanctions (Visciano & Schirone, 2021).

METHODS USED TO DETECT MILK FRAUD

The methods used to identify fraud and adulteration in milk can be classified into qualitative and quantitative (Ionescu et al., 2023). Qualitative methods are based on chemical reactions that result in color changes, encompassing biochemical and physicochemical processes. Although they have lower precision, they stand out for their simplicity, speed, and ease of application (Fanelli et al., 2021). On the other hand, quantitative methods involve more advanced instrumental techniques and sophisticated molecular tools, allowing for greater sensitivity, accuracy, and agility in the detection and analysis of adulterants present in dairy products, in addition to contributing to the authentication and traceability of these foods (Ionescu et al., 2023). The main differences and applications between the two methods are discussed in Table 1.

Table 1. Main differences and applications between qualitative and quantitative methods for milk fraud detection

Aspect	Qualitative Methods	Quantitative Methods
Detection Base	Chemical reactions	Instrumental techniques
Precision	Low	High
Rapidity	Fast	Usually quick (depends on technique)
Intricacy	Simple	High

Aspect	Qualitative Methods	Quantitative Methods
Application	Easy and affordable	Sophisticated equipment and specialized labor
Cost	Generally low	Generally high
Sensitivity	Limited	High
Purpose	Tamper Indication	Quantification and accurate identification of adulterants

The qualitative detection of adulterations in milk is based on observable physicochemical changes (Figure 2), and may result from the reaction between adulterants and specific reagents, accompanied by colorimetric changes, in this case, require a binary (yes/no) answer (Rahman et al., 2017). In addition, physicochemical properties such as pH, density, color, redox reactivity, rheology, conductivity, and thermal characteristics are sensitive to adulteration (Nagraik et al., 2021).

A classic example is the identification of starch in milk, which occurs through the reaction between starch and iodine, resulting in a characteristic blue coloration (Sharma et al., 2012). For the detection of urea, the reagent p-dimethylaminobenzaldehyde is used, whose reaction with the substance generates a specific yellow color (Singh et al., 2012). In the analysis to identify the presence of detergents or soap, a solution of bromocresol purple is used, which, in contact with these substances, produces a distinctive violet color (Singh et al., 2012). These analyses can be carried out in biosafety laboratories, provided that the necessary chemical reagents are available and appropriate safety measures are adopted to protect the professionals involved in the process (Azad & Ahmed, 2016).

Figure 2. Color-based reactions for the determination of adulterants in milk (adapted from Chaudhry et al., 2015)

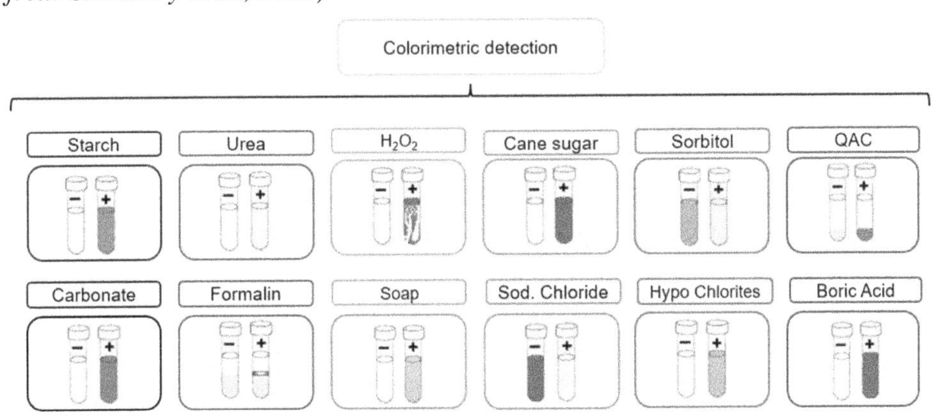

The main limitations of these methods lie in their low precision and the restriction of their applicability to a limited range of concentrations (Poghossian et al., 2019). To overcome these disadvantages, two main strategies have been implemented: the standardization of the chemical reagents used, aiming at greater uniformity in the results, and the optimization of the experimental conditions in which the tests are performed, providing greater reliability and efficiency in the procedures (Aparnathi et al., 2020).

Currently, several quantitative methods have been developed for the detection of fraud and adulterants in milk (Figure 3). These methods provide quantitative and qualitative data, while having high sensitivity, adequate turnaround time, and minimal sample preparation (Khosroshahi et al., 2022). Among the main analytical techniques used are spectroscopy, mass spectrometry, chromatography, electrokinetic separation and sensors. These techniques make it possible to identify specific components, quantify their levels in food, and detect compounds in food samples with complex chemical compositions (Danezis et al., 2016).

Although the chromatographic methods, High Performance Liquid Chromatography (HPLC) and Gas Chromatography (GC), are officially recognized for the detection of harmful substances in dairy products, they involve expensive equipment, long and laborious procedures, and analysis of results (Chen et al., 2004; Nagraik et al., 2021).

In this context, other techniques have been gaining ground because they have a non-destructive character of the samples (Shawky et al., 2024). On the other hand, it is worth noting that these non-destructive approaches to detect fraud and adulteration are often expensive and may become less effective due to the complexity of the food (Fiorino et al., 2018; Shawky et al., 2024).

Spectroscopy technology is based on the interaction between electromagnetic radiation and matter, varying as a function of wavelength, frequency, or energy. This broad range of radiation includes regions such as ultraviolet (UV), visible (VIS), near, mid, and far infrared (NIR, MIR, and FIR), as well as X-rays (Zhu et al., 2022; Frigerio et al., 2024).

Mass spectrometry is an analytical technique capable of providing qualitative and quantitative information about proteins and peptides present in food, based on the ionization of molecules and the classification of the ions generated according to their mass/charge ratio (m/z). Among the main types of mass spectrometry are isotope ratio mass spectrometry (IRMS); inductively coupled plasma mass spectrometry (ICP-MS); proton transfer reaction mass spectrometry (PTR-MS).

Electrokinetic separation techniques, which classify ions based on their electrophoretic mobility. Among these methods are capillary electrophoresis (CE), gel electrophoresis (SDS-PAGE), polymerase chain reaction (PCR), and simple sequence repeats (SSR), widely used to assess the authenticity of food (He et al.,

2021; Vinothkanna et al., 2024)element. In particular, CE and PCR stand out as the most used tools for this purpose (Vinothkanna et al., 2024). In the following section, the application of these technologies in the detection of fraud in milk will be discussed, focusing on their effectiveness and specificity to identify adulteration in this type of food.

Biosensors are analytical devices composed of a bio-receiver, transducer, and signal reader (Mascini & Tombelli, 2008). The bio-receptor can be formed by molecules such as enzymes, antibodies, or nucleic acids, while the transducer converts biological signals into electrical ones with high accuracy (Nagraik et al., 2021). Designed for maximum sensitivity and specificity, biosensors can be affinity, forming complexes with the analyte, or catalytic, reacting directly with it (Nagraik et al., 2021). Biosensors can be used for the detection of various milk adulterants, such as melamine, urea, non-dairy fat, proteins, preservatives, etc.

Figure 3. Quantitative methods for identifying fraud in milk.

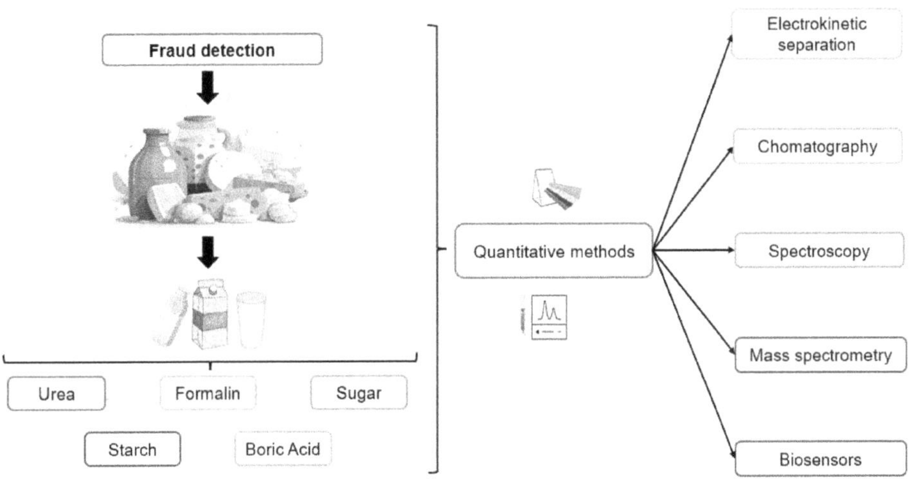

MODERN TECHNOLOGIES FOR MILK QUALITY MONITORING

Food authentication requires effective analytical methods to ensure accurate assessments and support safety management and monitoring systems at all stages of the milk production and chain (Spink, 2019). This process includes the analysis of compounds, and indicates whether the composition, purity, origin and processing methods of food are accurate and whether the content is authentic, comparing the results obtained with the standard values of genuine materials (Theodoridis & Kraemer, 2018).

In fact, there are several analytical techniques, each tailored to the specific food, its unique characteristics, and its complexity of composition. Advanced techniques, such as spectroscopy, mass spectrometry, chromatography, and electroanalytical methods, have been widely studied to identify fraud and verify the authenticity of milk (Table 2) (Ferreira et al., 2024).

Table 2. Instrumental techniques for detecting fraud and adulteration in milk

Techniques	Parameters	Fraud/ Tampering	Findings	References
Spectroscopy				
NIR	1070–980 cm $^{-1}$	Addition of Melamine in Cow's Milk	The combination of NIR and OCPLS spectroscopy detected the presence of melamine in milk samples with total accuracy of 89%, sensitivity of 90%, and specificity of 88%.	Chen et al. (2017)
MIR and Raman	MIR: 1373, 1454 and 956 cm $^{-1}$ Raman: 1005, 1154 and 1551 cm $^{-1}$	Adding cow's milk to goat's milk	Both spectroscopic methods accurately identified the percentage of enriched milk	Yaman (2020)
MIR	4000–650 cm $^{-1}$	Addition of cow's milk to buffalo and goat milks	The method distinguished the mixtures of goat-cow milk and buffalo-cow milk with 93% and 91% correct classification rates	Sen et al. (2021)
Fluorescence	295 nm	Adding whey to genuine milk	The method allowed the discrimination between genuine and adulterated milk, however, it was not possible to classify the samples according to the level of adulteration	Lelis et al. (2022)
MIR and Fluorescence	MIR: 2500 and 650 cm $^{-1}$ Fluorescence: 20, 40, 60, 80, 100 and 120 nm	Addition of cow, sheep and goat milk in camel milk	Fluorescence spectroscopy was the most accurate and sensitive technique to identify and predict the level of camel tampering.	Boukria et al; (2023)
MIR	925 to 5011 cm $^{-1}$	Adding cow's milk to goat's milk	The MIR method obtained optimal classification and excellent quantitative accuracy for goat's milk adulteration	Du et al. (2025)

Techniques	Parameters	Fraud/ Tampering	Findings	References
Raman	2300 to 400 cm^{-1}	Adding Cow's Milk Powder to Goat Milk Powder	Raman spectroscopy allowed the authentication of pure goat's milk powder, with a significance level of 0.01.	Ferreira et al. (2025)
Mass spectrometry				
Nano-HPLC-MS/MS (TANDEM Mass Spectroscopy)		Adulteration of milk with vegetable proteins	The nano-HPLC-MS/MS indicated the main proteins of soybean (β-conglycinin, glycinin), peas (vinciline, convicilin, legume) and wheat (glutenin and gliadin) in adulterated milk	Yang et al. (2018)
Mass spectrometry (UPLC-QTOF-MS)	50 ~1,500 m/z	Adding powdered milk to whole liquid milk	The methods efficiently differentiated adulterated liquid whole milk from powdered milk	Du et al. (2020)
Ionization Mass Spectrometry (MALDI-MS)	5000–30,000 m/z	Addition of bovine milk to goat and sheep milk	The MALDI-MS method created the profile of the main proteins allowing to quantify the adulteration of goat and sheep milk with cow's milk	Tehrani et al. (2024)
Capillary isoelectric mass spectrometry	500–4000 m/z	Identification of whey proteins in dairy products	The methodology allowed to accurately identify intact whey proteins in dairy products	Chan et al. (2025)
Electrokinetic separation				
Capillary electrophoresis	214 nm	Addition of bovine milk in pasteurized horse milk and horse milk powder	Capillary electrophoresis allowed the identification of several bovine proteins and metabolites in pasteurized milk and mare's milk powder	Ji et al. (2024)
Direct PCR with fluorescence	460 nm	Detection of cow DNA in milk and dairy products	The method was able to identify fraudulent substitutions in non-bovine milk and dairy products	Jitsuwan et al. (2024)
Multiplex PCR	260 and 280 nm	Detection of camel milk with bovine and goat milk	Multiplex PCR showed detection limits of 5% and 10% for goat milk and cattle milk in camel milk adulteration	Sarkar et al. (2024)
Biosensors				
Electrochemical biosensor	Detection limit 13h to 15h	Adding melamine and urea to milk	The biosensor detected melamine and urea in a range of 1–20 nM with a detection limit of 3 pM and 1 pM, respectively	Ezhilan et al. (2017)
Optical Biosensor	N/A	Addition of hydrogen peroxide (H2O2) in milk	The biosensor detected adulteration in milk with hydrogen peroxide at concentrations between 0.002% w/w and 0.006%	Vasconcelos et al. (2021)
Silver Nanostructures (AgNS) Sensor	Melamine: 1.69×10^{3} Urea: 480	Adding melamine and urea to milk	The sensor detected the presence of melamine and urea with sensitivity of 1.69 and 0.48 mM	Verma et al. (2021)

continued on following page

Table 2. Continued

Techniques	Parameters	Fraud/ Tampering	Findings	References
Colorimetric sensor of zein/MnO 2 nanosheets	N/A	Addition of hydrogen peroxide (H2O2) and lactic acid	The sensor detected adequate presence of $_{H2O2}$ and lactic acid of 7.2 $\times 10^{-4}$ mol L^{-1} and 7.5 $\times 10^{-4}$ mol L^{-1}, respectively.	Pereira et al. (2024)
Electrochemical sensor	Potential range from – 0.2 V to 1.5 V	Adding cow's milk to goat's milk	The sensor identified the presence of cow's milk in goat's milk with more than 94% accuracy	Xue & Zhao (2024)
Fiber Optic Sensor	Deposition pressure 25 x 10^{-3} mbar, base pressure 3 x 10^{-5} mbar	Adding formaldehyde and water to milk	The sensor identified the milk fraud through changes in the resonance wavelength for different analytes	Harini et al. (2024)

Spectroscopy

Spectroscopic techniques have stood out for their high accuracy and reliability in the analysis of different types of biological samples, in addition to offering low cost for milk analysis (Ulla et al., 2020). Among the widely used technologies, near infrared (NIR), mid-infrared (MIR), Raman spectroscopy, fluorescence and nuclear magnetic resonance spectroscopy (NMR) stand out. These analyses, in addition to being fast, require minimal sample processing and a safer approach make spectroscopic methods appropriate for the detection of milk adulterants (Nagraik et al., 2021).

Melamine is often added to milk as an adulterant to raise the apparent protein content, due to its high nitrogen content. Its addition in milk can result in side effects that cause different diseases, such as reproductive damage, kidney stones, and bladder stones (Nagraik et al., 2021). In a study conducted by Chen et al. (2017), NIR spectroscopy was used in combination with OCPLS (partial least squares of a class) to detect the presence of melamine in milk samples. The model identified the presence of melamine in the milk samples, and achieved accuracy of 89%, sensitivity of 90% and specificity of 88% in the test set.

The adulteration of high-value milks, such as goat and buffalo milks, by adding lower-value milks, such as cow's milk, is one of the most common fraudulent practices in the dairy industry (Ionescu et al., 2023). Studies based on spectroscopic techniques have demonstrated efficacy in the identification and classification of adulterated milk in relation to genuine milk. Table 2 presents several studies in which techniques such as MIR, NIR, and fluorescence were effective in detecting cow's milk mixed with higher value-added milk, such as goat and buffalo milk (Yaman, 2020; Sen et al., 2021; Boukria et al., 2023; Du et al., 2025). These techniques have also been shown to be efficient in identifying the addition of whey to genuine milk (Lelis et al., 2022).

Mass Spectrometry

Mass spectrometry approaches have been widely applied to food authentication and traceability, providing valuable information about the geographical, chemical, and biological origins of food (Katerinopoulou et al., 2020). Among the mass spectrometry methodologies used in food analysis, those focused on protein and peptide analysis are leaders in evaluating the authenticity of foods (Valletta et al., 2021). These technologies stand out for their sensitivity, specificity, ability to perform multiplexing, and high throughput. In addition, the relative stability of proteins and peptides during food processing contributes to authenticity monitoring, increasing the efficiency and reliability of analyses (Valletta et al., 2021).

In a study developed by Chan et al. (2025), mass spectrometry was used to characterize whey protein in dairy products. The main advantage of this method is its ability to provide a fast and accurate analysis of serum proteins such as α-lactalbumin, β-lactoglobulin A, and B in a short period. High-resolution mass spectrometry has enabled accurate identification of proteins, significantly enhancing dairy analysis (Chan et al., 2025).

The combination of technologies can also be an effective strategy to detect adulteration in milk, such as the addition of other constituents. An example of this is the use of high-performance liquid chromatography (HPLC) associated with mass spectrometry (MS), which allows the identification of peptides from vegetable proteins added to milk (Luykx et al., 2007). Vegetable proteins are considered an economical alternative to increase dairy production. However, the presence of allergens coming from these proteins can cause severe reactions, such as anaphylaxis, posing a significant risk to consumer health (Nakamura & Teshima, 2013). In a study conducted by Yang et al. (2018), non-directed nano-high-performance liquid chromatography-tandem mass spectroscopy was used for detection of plant protein in adulterated milk. The results indicated that nano-HPLC-MS/MS was able to identify soybean, pea and wheat proteins, with the effective detection of soybean and wheat proteins from 0.5% of the total protein, and pea proteins at levels of 2% to 4%.

Electrokinetic Separation

Electrokinetic separation methods are analytical techniques that use an electric field to separate components in a sample based on their electrical properties, such as charge and size. These methods are quite effective in detecting fraud in food, including milk, as they allow the separation and identification of adulterating compounds or contaminants present, often in very low concentrations. Among them are capillary electrophoresis and polymerase chain reaction (PCR).

Capillary electrophoresis (CE) is a highly efficient analytical technique characterized by fast analysis times and low sample and solvent consumption (Wu & Zhang, 2013). Although it has lower sensitivity compared to methods that use mass spectrometry, CE has been widely applied in the detection of fraud in milk and dairy products. In a study conducted by Ji et al. (2024), capillary electrophoresis was used to evaluate samples of pasteurized horse milk and horse milk powder adulterated with cow's milk. The results showed that bovine milk constituents, such as osteopontin (OPN) and serotransferrin (TF), were able to detect adulteration of up to 1% in mare's milk, while the presence of biotin and N6-Me-adenosine proved effective in identifying adulteration at levels of up to 10% and 1%, respectively.

Polymerase chain reaction (PCR) is a widely used technique for the simultaneous identification and amplification of multiple target sequences. Its applications include pathogen detection, genetic screening, and gene expression analysis (Sarkar et al., 2024). In view of the recurrent practice of adulteration of milk of different species, Sarkar et al. (2024) conducted a study to identify adulterations in camel milk with bovine and goat milk, using the multiplex PCR technique. The method allowed the detection and authentication of adulterated camel milk by means of specific DNA fragments (208 bp for camel, 274 bp for cattle and 174 bp for goat). The detection limit was 10% for bovine milk and 5% for goat milk in camel milk, demonstrating the efficacy and precision of the technique.

Biosensors

Biosensors can be divided into different types, which are colorimetric, luminescent, surface plasmonic resonance (SPR) and fiber optics. Among them, colorimetric and fluorescent approaches are widely used due to their ease of manufacture, low cost, simplicity, and straightforward results (Shalileh et al., 2023). Colorimetric biosensors detect analytes through color changes, visible to the naked eye or with simple devices such as smartphones (Zhao et al., 2020). Luminescent light, on the other hand, uses changes in the fluorescent light emitted to indicate the concentration of the analyte. Electrochemists, which combine biological components for specific detection, stand out for their advantages such as low detection limits, easy miniaturization, low cost, and the ability to analyze turbid samples without the need for pretreatment (Shalileh et al., 2023).

Colorimetric sensors offer numerous advantages for detecting fraud in milk, as they have high sensitivity, cost-effectiveness, ease of use, fast analysis, direct operation, and clear visibility to the naked eye (Pereira et al., 2024). In a study developed by Pereira et al. (2024), a colorimetric sensor of zein/MnO2 nanosheet composites was developed for the identification of two milk contaminants, hydrogen peroxide (H2O2) and lactic acid. The sensor showed good sensitivity and detection limits of 7.2×10^{-4} mol L^{-1} for H_2O_2 and 7.5×10^{-4} mol L^{-1} for lactic acid, within the limits established by international regulatory bodies. In addition, the detection was carried out on economical and ecological substrates, such as filter paper and adhesive tape, using a simple airbrush, being a sustainable alternative for fraud detection in milk.

Biosensors are widely used for detecting melamine in milk. An electrochemical biosensor based on Pt electrode modified with zinc oxide nanospheres was developed for the simultaneous determination of melamine and urea in cow's milk sample (Chen et al., 2017). The developed biosensor demonstrated high accuracy and selectivity in the simultaneous detection of melamine and urea in cow's milk, with extremely low detection limits (3 pM for melamine and 1 pM for urea). The linear regression models used for analysis showed high efficacy, with recovery between 99.96% and 102.22%.

The addition of H2O2 is widely used in the dairy industry as an antimicrobial agent, thus aiding in the preservation of raw milk in the absence of refrigeration. In a study by Vasconcelos et al. (2021), a chemiluminescence-based optical biosensor was developed, using a low-cost sensitive hydroxyethylcellulose membrane combined with a high-sensitivity photodetector to measure H_2O_2 concentrations in semi-skimmed milk. The method showed good repeatability and was able to detect concentrations as low as 0.001% w/w. Adulterated milk analysis revealed concentrations between 0.002% w/w and 0.006%, and the method was calibrated for on-site analysis, with an adequate detection limit.

Electronic Nose and Tongue in Milk Fraud Control

The electronic nose and electronic tongue were fast, non-destructive and cost-effective techniques for food authentication and have become alternatives to traditional laboratory techniques (Figure 4) (Titova & Nachev, 2020).

Electronic noses (E-noses) detect and distinguish complex odors through non-specific sensors treated with odor-sensitive substances (Peris & Escuder-Gilabert, 2016). These sensors generate characteristic patterns that, when analyzed, allow the identification and classification of odors. Technological advances have made it possible to use gas sensors and mass spectrometry for greater precision. The basic system of these devices includes sample handling, detection and data processing (Banti, 2020). Electrochemical sensors operate with interactions between gas molecules and coating materials, converting signals into digital data processed by microprocessors for statistical analysis (Mahanti et al., 2024). There are several types of sensors, classified by materials, chemical interactions, and transduction methods, that can measure electrical properties, mass changes, temperature, or optical characteristics, such as fluorescence and light absorbance (Mahanti et al., 2024; Peris & Escuder-Gilabert, 2016).

Electronic tongues, similar to electronic noses, identify patterns of chemical substances responsible for flavor, processing the data statistically (Peris & Escuder-Gilabert, 2016; Imam & Nagpal, 2023). Composed of various chemical sensors, they usually include an automatic sampler, a set of sensors with moderate selectivity, and software to process signals (Facure et al., 2023). The system uses sensors with cross-sensitivity, allowing each one to provide information on several analytes. Sensor signals are analyzed by chemometric methods, such as principal component analysis, which helps characterize samples in terms of analyte concentration and identify their natures (Yashaswini et al., 2022).

Figure 4. Operating mechanism and important components of the E-nose and E-tongue system

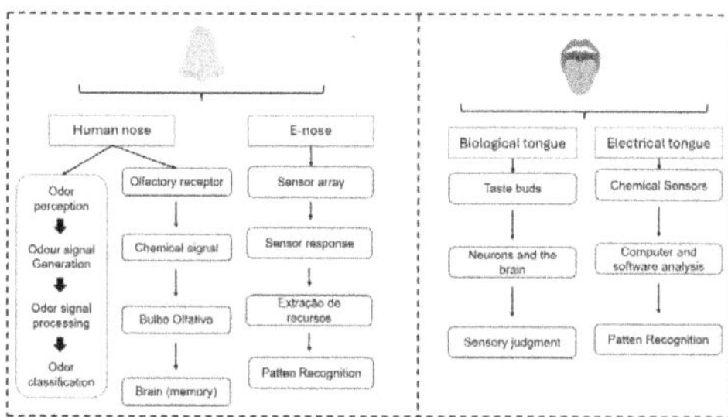

The integration of E-nose and E-tongue technologies at various stages of the food supply chain ensures food safety. They are emerging as promising complementary techniques to classical analytical methods for rapid and low-cost detection of fraud in food products, such as dairy products (Bueno et al., 2014; Majcher et al., 2015; Ayari et al., 2018; Tian et al., 2023; Arrieta et al., 2023).

Studies have demonstrated the effectiveness of electronic tongue (E-tongue) in identifying adulterations in milk. In one experiment, E-tongue detected urea, formaldehyde and melamine at concentrations of 10.0, 4.16 and 0.95 mmol/L, respectively, using PCA-based analyses (Bueno et al., 2014). Furthermore, adulteration of goat milk with cow milk was identified with 100% accuracy using kernel principal component analysis (KPCA) for qualitative discrimination (Han et al., 2019). For quantitative assessment, extreme machine learning (ELM) algorithms were applied, highlighting the potential of E-tongue in food authentication and quality control.

Despite innovations, electronic noses and tongues have limitations, such as low sensitivity and selectivity, cross-sensitivity, and the impact of environmental variables, such as temperature and humidity, which can compromise accuracy. Maintenance, including frequent calibration and replacement of sensors, increases operating costs.

These challenges limit their usefulness in some applications, but research is seeking to improve stability, selectivity, and accessibility (Mahanti et al., 2024).

Challenges include sensor drift and lack of specificity to differentiate similar compounds, causing inaccuracies in analysis. New materials, such as nanomaterials and bioinspired compounds, offer promising advances, increasing sensitivity and stability. Additionally, the integration of machine learning and artificial intelligence enhances data analysis, improving the detection of patterns and adulterations in food (Mahanti et al., 2024).

CHEMOMETRICS TO DETECT MILK ADULTERATION

Chemometrics is the science that uses mathematical and statistical methods to optimize the processing of food data, being essential in the authentication and detection of product adulteration (Figure 5) (Grassi et al., 2023). Chemometric tools, such as multivariate analysis, allow for the identification of patterns in large data sets and the development of calibration models (Capuano et al., 2014). Principal component analysis (PCA) and hierarchical cluster analysis (HCA) are examples of techniques used to reduce variables and group samples based on their properties, and are most often employed for food analysis (Roggo et al., 2007).

Figure 5. Chemometric analysis workflow

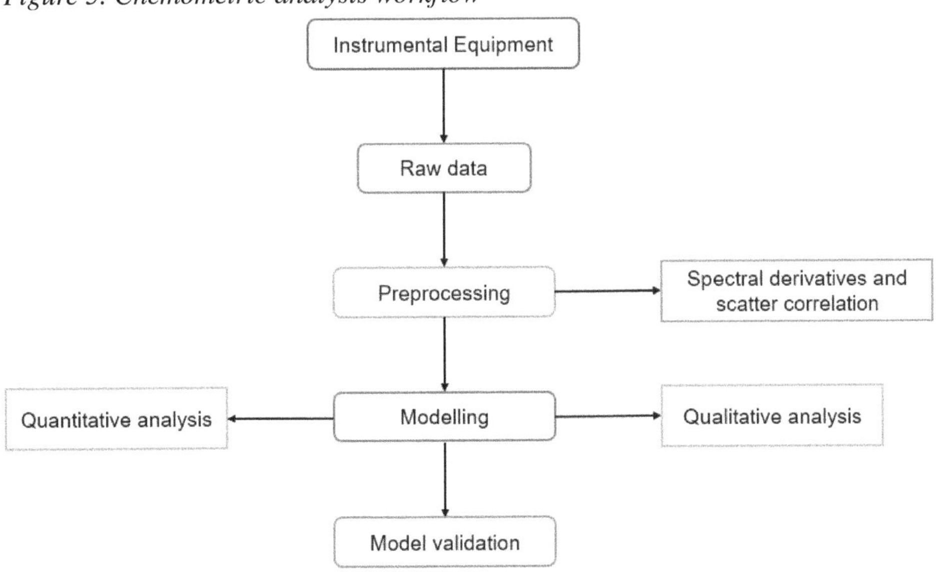

One of the main factors that highlight the importance of chemometric techniques (mathematical and statistical methods) in data manipulation is the inherent complexity of food (Moghaddam et al., 2022). In the case of milk, physical properties such as particle size often generate challenges, including variations in optical path length, light scattering, overlapping spectral bands, and changes in baseline (Moghaddam et al., 2022). These characteristics make it essential to pre-process the data to ensure that a reliable and accurate calibration model is obtained (Cen & He, 2007).

Although it is a powerful analysis, chemometrics are especially more effective when combined with established investigative methods. Its effectiveness also depends on the availability of reliable reference data for authentic products and potential adulterants (Sharma et al., 2024). Techniques such as regression and pattern recognition are widely applied to assess the authenticity of food, enabling qualitative and quantitative analyses through multivariate methods (Sharma et al., 2024).

Table 3 presents studies involving the application of different chemometric techniques to various analytical data covering previously discussed technologies to combat milk adulteration.

Table 3. Recent studies related to the adulteration and authenticity of milk in combination with chemometric analysis

Type of milk	Type of fraud	Analytical method	Reference
Cow, sheep and buffalo milk	Addition of milk mixture from different species	FTIR	Cirak et al. (2018)
Milk powder	Addition of melamine	Raman	Karunathilaka et al. (2018)
UTH (skimmed and semi-skimmed) milk and raw milk	Adding Cheese Whey	FTIR-ATR	Vinciguerra et al. (2019)
Cow's milk	Adding water	NIR	Kamboj et al. (2020)
Cow's milk	Addition of urea	NIR	Mabood et al. (2020)
Cow's milk	Addition of formaldehyde, based on aldehydes and ketones	Colorimetric sensors	Mostafapour et al. (2021)
Cow's milk	Addition of maltodextrin, sodium carbonate, whey	Raman	Tian et al. (2022)
Cow's milk	Addition of melamine	FT-IR	Jin et al. (2023)
Mare's milk	Addition of goat's milk	LC-HRMS	Windarsih et al. (2024)
Milk powder	Addition of cornstarch and wheat flour	FT-NIR and FT-MIR.	Caballero-Agosto et al. (2024)
Goat's milk powder	Addition of powdered cow's milk	Raman	Ferreira et al. (2025)
Goat milk	Addition of cow's milk	MIR	Du et al. (2025)

The study by Jin et al. (2023) presents an advanced approach to detecting adulteration in milk using two-dimensional autocorrelation spectroscopy combined with chemometric methods such as least-squares support vector machines (LS-SVM). The multivariate analysis allowed a significant reduction in the variables analyzed, identifying adulterations with high accuracy (up to 99.05% for multiple brands). This combination has proven to be a robust tool for discriminating against melamine-adulterated milk, optimizing the interpretation of spectroscopic data and showing potential to ensure food authenticity and safety.

In another study, Ferreira et al. (2025), used Raman spectroscopy in combination with chemometric techniques to authenticate powdered goat's milk adulterated with powdered cow's milk. Models based on DD-SIMCA have been developed to distinguish pure and adulterated samples with high accuracy. Data preprocessing, such as multiplicative scattering correction (MSC) and the Savitzky-Golay derivative, stood out in improving the quality of spectral analyses. These models showed sensitivity and specificity of 100%, validating the effectiveness of the approach to detect tampering at levels as low as 1%.

Finally, Du et al. (2024) discussed the combined use of ATR-FTIR spectroscopy and chemometrics for adulteration detection in milk. Chemometric analysis is essential for interpreting complex spectroscopic data, especially in mixtures such as raw milk adulterated with milk powder. The SVM and PLS-DA models showed high accuracy (99%) in classifying adulterated samples, highlighting the effectiveness of these approaches compared to traditional methods. The application of these techniques offers a fast, sensitive and non-destructive solution, expanding the possibilities of continuous monitoring and quality control in the dairy industry.

BLOCKCHAIN TECHNOLOGY TO ASSIST IN COMBATING MILK FRAUD

Despite the significant advantages of the previously mentioned techniques for detecting fraud and adulteration in the milk production chain, the increasing complexity of global food supply chains and the frequent cases of adulteration highlight the need to develop even more robust and efficient methods for food authentication (Bhat et al., 2025). In this context, the integration of smart technologies has become an essential transformation. Tools such as blockchain technology have been bringing remarkable innovations to this area (Goyal et al., 2022).

Blockchain technology has gained prominence for its ability to ensure greater transparency, traceability, and accountability in food supply chains, minimizing fraud and increasing consumer trust (Figure 6) (Duan et al., 2024). In the food industry, this technology enables real-time monitoring of all stages of the production chain, using sensors, RFID tags, and smart contracts to record information validated by consensus among participants. This allows you to quickly trace the source of contaminations during crises, as well as verify the authenticity and quality of premium and organic products (Goyal et al., 2022). This capability makes blockchain a valuable tool in the fight against food fraud. The technology is incorporated into the food system through unique digital labels for each product, which are monitored by smart contracts and IoT sensors throughout the supply chain, creating a transparent and traceable path, which decreases the chance of fraud going unnoticed (Vern et al., 2024).

Figure 6. Blockchain innovations in milk fraud prevention

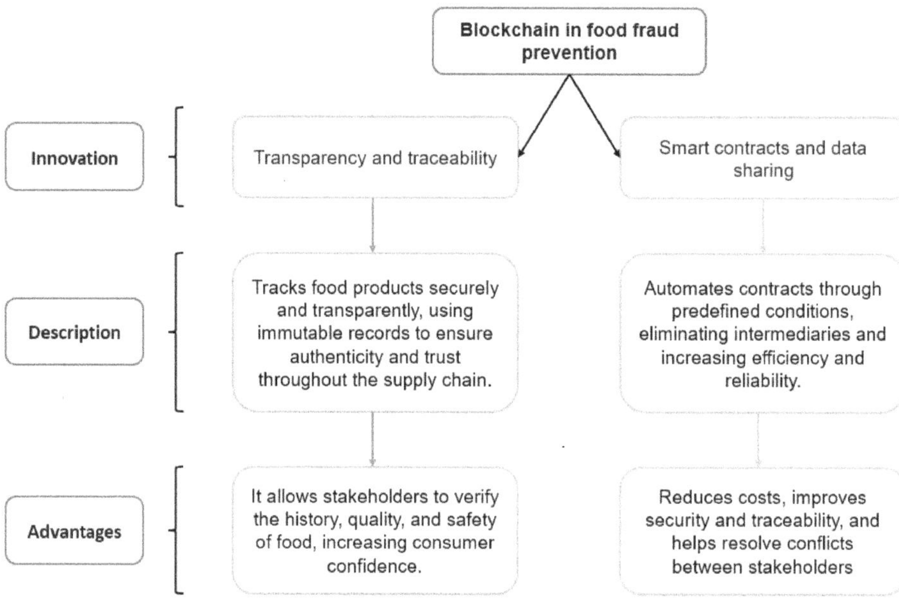

On the other hand, the implementation of blockchain in the fight against food fraud still faces some challenges, such as scalability, interoperability, and data accuracy (Whyte et al., 2022). Scalability is a growing issue as the blockchain network expands, with the increased number of transactions leading to slowdowns and higher costs (Patel et al., 2023). Additionally, interoperability between blockchain platforms is hampered by the lack of common standards, making it difficult to integrate data into complex supply chains, such as food (Mohammed et al., 2023). Data accuracy and privacy are also crucial, as errors can compromise the system, and sharing sensitive information can raise security concerns (Whyte et al., 2022).

Despite these challenges, blockchain technology has the potential to enhance the safety and quality of food supply chains by offering traceability and accountability. For this, it is necessary to advance solutions through technological innovations, such as IoT and Artificial Intelligence, and the adoption of changes in the supply chain to ensure greater safety and health for consumers (Duan et al., 2024).

FUTURE PERSPECTIVES IN MILK QUALITY MONITORING

Food adulteration, characterized by the intentional addition or subtraction of substances, compromises the quality and, in some cases, the nutritional properties of food (Ferreira et al., 2024). In the case of milk, it is common to dilute it with water, add solids to compensate for low quality, include proteins or chemicals to increase nutritional value, or disguise lower-value milk as a premium product (Ferreira et al., 2024). This scenario underscores the importance of controlling quality to meet consumer demands and ensure food integrity (Qin et al., 2022).

The future prospects for the use of instrumental techniques in milk quality monitoring are promising, driven by technological advances and the growing demand for high-quality and safe dairy products (Wadood et al., 2022). Instrumental techniques are being integrated into automated systems to continuously monitor milk fraud. These techniques allow for quick and accurate analysis of milk quality without changing its composition, offering significant benefits by preserving milk integrity and ensuring quality control (Karabagiaset al., 2024).

The future prospects for the use of instrumental techniques in milk quality monitoring are being enhanced with the integration of emerging technologies such as blockchain technology (Duan et al., 2024). With the increasing demand for high-quality and safe dairy products, automation and real-time monitoring are gaining increasing relevance (Whyte et al., 2022).Instrumental techniques such as sensors and spectrometers are incorporated into automated systems to continuously monitor essential milk parameters, providing immediate feedback to producers and processors (Mohammed et al., 2023).

In addition, the combination of instrumental techniques with chemometrics is also providing more accurate and sophisticated interpretations of the data. This can result in more accurate predictive models for milk quality by identifying potential problems based on subtle variations in the chemical composition of milk (Moghaddam et al., 2022). Additionally, as the demand for sustainable practices increases, instrumental techniques will play a key role in optimizing milk production processes. By improving the accuracy of quality assessments, these techniques can reduce waste, ensure better resource management, and enhance the sustainability of the dairy industry (Sharma et al., 2024).

CONCLUSION

Monitoring milk quality is essential to ensure global food security, considering its relevance in the human diet and its susceptibility to fraud and contamination. While technological advances in the dairy sector have increased production efficiency, challenges related to adulteration, microbiological contamination, and chemical changes along the supply chain remain significant. In this context, the use of modern sensors, such as those based on spectroscopy, biosensors, and electrochemical sensors, combined with chemometric tools, provides more robust and accurate analyses, facilitating the detection of adulterants and maintaining milk quality.

Additionally, blockchain technology emerges as an innovative solution to track and document all stages of the production chain, promoting greater transparency and reliability. This technology ensures that critical information about the origin, transport, and storage of milk is recorded in an immutable and accessible manner, reducing the risks of adulteration and increasing consumer confidence.

While the initial implementation costs and technical barriers are challenges to its large-scale adoption, the long-term benefits, such as improved food safety, reduced losses, and increased operational efficiency, justify the investment. In addition, the integration of these technologies with big data and artificial intelligence promises even more significant advances in predictive analytics and decision-making. Therefore, the integration of modern sensors, chemometric techniques, and blockchain technology represents a consolidated and indispensable approach for ensuring milk quality, meeting regulatory standards, and addressing consumer expectations for safety and authenticity.

REFERENCES

Afzal, A., Mahmood, M. S., Hussain, I., & Akhtar, M. (2011). Adulteration and microbiological quality of milk (a review). *Pakistan Journal of Nutrition*, *10*(12), 1195–1202. DOI: 10.3923/pjn.2011.1195.1202

Aparnathi, K. D., Shaikh, A. I., & Patel, S. I. (2020). *Qualitative Tests for Detection of Common Adulteration in Milk. Directorate of Research Anand Agricultural University*. Anand.

Arrieta, A., Barrera, I., & Mendoza, J. (2023). Application of a Polypyrrole Sensor Array Integrated into a Smart Electronic Tongue for the Discrimination of Milk Adulterated with Sucrose. *International Journal of Technology*, *14*(1), 90–99. DOI: 10.14716/ijtech.v14i1.5613

Ayari, F., Mirzaee-Ghaleh, E., Rabbani, H., & Heidarbeigi, K. (2018). Detection of the adulteration in pure cow ghee by electronic nose method (case study: Sunflower oil and cow body fat). *International Journal of Food Properties*, *21*(1), 1670–1679. DOI: 10.1080/10942912.2018.1505755

Azad, T., & Ahmed, S. (2016). Common milk adulteration and their detection techniques. *International Journal of Food Contamination*, *3*(1), 1–9. DOI: 10.1186/s40550-016-0045-3

Banti, M. (2020). Food adulteration and some methods of detection [review]. *International Journal of Nutrition and Food Sciences*, *9*(3), 86–94. DOI: 10.11648/j.ijnfs.20200903.13

Bhat, M. A., Rather, M. Y., Singh, P., Hassan, S., & Hussain, N. (2024). Advances in Smart Food Authentication for Enhanced Safety and Quality. *Trends in Food Science & Technology*, 104800.

Boukria, O., Boudalia, S., Bhat, Z. F., Hassoun, A., & Aït-Kaddour, A. (2023). Evaluation of the adulteration of camel milk by non-camel milk using multispectral image, fluorescence and infrared spectroscopy. *Spectrochimica Acta. Part A: Molecular and Biomolecular Spectroscopy*, *300*, 122932. DOI: 10.1016/j.saa.2023.122932 PMID: 37270971

Brooks, C., Parr, L., Smith, J. M., Buchanan, D., Snioch, D., & Hebishy, E. (2021). A review of food fraud and food authenticity across the food supply chain, with an examination of the impact of the COVID-19 pandemic and Brexit on food industry. *Food Control*, *130*, 108171. DOI: 10.1016/j.foodcont.2021.108171

Bueno, L., De Araujo, W. R., Salles, M. O., Kussuda, M. Y., & Paixão, T. R. (2014). Voltammetric electronic tongue for discrimination of milk adulterated with urea, formaldehyde and melamine. *Chemosensors (Basel, Switzerland)*, *2*(4), 251–266. DOI: 10.3390/chemosensors2040251

Caballero-Agosto, E. R., Sierra-Vega, N. O., Rolon-Ocasio, Y., Hernandez-Rivera, S. P., Infante-Degró, R. A., Fontalvo-Gomez, M., Pacheco-Londoño, L. C., & Infante-Castillo, R. (2024). Detection and quantification of corn starch and wheat flour as adulterants in milk powder by near- and mid-infrared spectroscopy coupled with chemometric routines. *Food Chemistry Advances*, *4*, 100582. DOI: 10.1016/j.focha.2023.100582

Cen, H., & He, Y. (2007). Theory and application of near infrared reflectance spectroscopy in determination of food quality. *Trends in Food Science & Technology*, *18*(2), 72–83. DOI: 10.1016/j.tifs.2006.09.003

Chan, S. L., Kwok, T., Xu, N., Bo, T., & Huang, T. (2025). Imaged capillary isoelectric focusing and online mass spectrometry for milk whey protein characterization in dairy products. *Analytical Biochemistry*, *699*, 115765. DOI: 10.1016/j.ab.2025.115765 PMID: 39778730

Chaudhry, H. R., Muhammad, K., & Rabbani, M. (2015). Laboratory manual quality control of milk: quality control of milk. *Amazon2015 Aug*.

Chen, H., Tan, C., Lin, Z., & Wu, T. (2017, February 15). Detection of melamine adulteration in milk by near-infrared spectroscopy and one-class partial least squares. *Spectrochimica Acta. Part A: Molecular and Biomolecular Spectroscopy*, *173*, 832–836. DOI: 10.1016/j.saa.2016.10.051 PMID: 27816741

Chen, R. K., Chang, L. W., Chung, Y. Y., Lee, M. H., & Ling, Y. C. (2004). Quantification of cow milk adulteration in goat milk using high-performance liquid chromatography with electrospray ionization mass spectrometry. *Rapid Communications in Mass Spectrometry*, *18*(10), 1167–1171. DOI: 10.1002/rcm.1460 PMID: 15150843

Cirak, O., Icyer, N. C., & Durak, M. Z. (2018). Rapid detection of adulteration of milks from different species using Fourier Transform Infrared Spectroscopy (FTIR). *The Journal of Dairy Research*, *85*(2), 222–225. DOI: 10.1017/S0022029918000201 PMID: 29785908

Danezis, G. P., Tsagkaris, A. S., Brusic, V., & Georgiou, C. A. (2016). Food authentication: State of the art and prospects. *Current Opinion in Food Science*, *10*, 22–31. DOI: 10.1016/j.cofs.2016.07.003

Deng, Z., Wang, T., Zheng, Y., Zhang, W., & Yun, Y. H. (2024). Deep learning in food authenticity: Recent advances and future trends. *Trends in Food Science & Technology*, *144*, 104344. DOI: 10.1016/j.tifs.2024.104344

Du, C., Zhao, X., Chu, C., Nan, L., Ren, X., Yan, L., Zhang, X. J., Zhang, S. J., & Teng, Z. (2025). Identification and quantification of goat milk adulteration using mid-infrared spectroscopy and chemometrics. *Spectrochimica Acta. Part A: Molecular and Biomolecular Spectroscopy*, *324*, 124969. DOI: 10.1016/j.saa.2024.124969 PMID: 39153347

Du, L., Lu, W., Zhang, Y., Gao, B., & Yu, L. (2020). Detection of milk powder in liquid whole milk using hydrolyzed peptide and intact protein mass spectral fingerprints coupled with data fusion technologies. *Food Science & Nutrition*, *8*(3), 1471–1479. DOI: 10.1002/fsn3.1430 PMID: 32180956

Duan, K., Onyeaka, H., & Pang, G. (2024). Leveraging blockchain to tackle food fraud: Innovations and obstacles. *Journal of Agriculture and Food Research*, *18*, 101429. DOI: 10.1016/j.jafr.2024.101429

Facure, M. H., Braunger, M. L., Mercante, L. A., Paterno, L. G., Riul, A., & Correa, D. S. (2022). Electrical Impedance-Based Electronic Tongues. *Encyclopedia of Sensors and Biosensors: Volume 1-4,* 567-590.

Fanelli, V., Mascio, I., Miazzi, M. M., Savoia, M. A., De Giovanni, C., & Montemurro, C. (2021). Molecular approaches to agri-food traceability and authentication: An updated review. *Foods*, *10*(07), 1644. DOI: 10.3390/foods10071644 PMID: 34359514

Fernando, I., Fei, J., Cahoon, S., & Close, D. C. (2024). A review of the emerging technologies and systems to mitigate food fraud in supply chains. *Critical Reviews in Food Science and Nutrition*, •••, 1–28. DOI: 10.1080/10408398.2024.2405840 PMID: 39356551

Ferreira, J. L. A., de Almeida, L. F., da Silva Simões, S., Diniz, P. H. G. D., & de Sousa Fernandes, D. D. (2025). Raman spectroscopy-based authentication of powder goat milk adulteration with cow milk. *Food Control*, *167*, 110800. DOI: 10.1016/j.foodcont.2024.110800

Ferreira, M. M., Marins-Gonçalves, L., & De Souza, D. (2024). An integrative review of analytical techniques used in food authentication: A detailed description for milk and dairy products. *Food Chemistry*, *457*, 140206. DOI: 10.1016/j.foodchem.2024.140206 PMID: 38936134

Fiorino, G. M., Garino, C., Arlorio, M., Logrieco, A. F., Losito, I., & Monaci, L. (2018). Overview on untargeted methods to combat food frauds: A focus on fishery products. *Journal of Food Quality*, *2018*(1), 1581746. DOI: 10.1155/2018/1581746

Frigerio, J., Campone, L., Giustra, M. D., Buzzelli, M., Piccoli, F., Galimberti, A., Cannavacciuolo, C., Ouled Larbi, M., Colombo, M., Ciocca, G., & Labra, M. (2024). Convergent technologies to tackle challenges of modern food authentication. *Heliyon*, *10*(11), e32297. DOI: 10.1016/j.heliyon.2024.e32297 PMID: 38947432

Goyal, K., Kumar, P., & Verma, K. (2022). Food adulteration detection using artificial intelligence: A systematic review. *Archives of Computational Methods in Engineering*, *29*(1), 397–426. DOI: 10.1007/s11831-021-09600-y

Haldar, L., Raghu, H. V., & Ray, P. R. (2022). Milk and milk product safety and quality assurance for achieving better public health outcomes. *Agriculture. Livestock Production and Aquaculture: Advances for Smallholder Farming Systems*, *1*, 217–259. DOI: 10.1007/978-3-030-93258-9_13

Han, H., Wang, Z., Li, C., Ma, Z., Yang, Z., & Ma, X. (2019). Purity detection of goat milk based on electronic tongue and improved artificial fish swarm optimized extreme learning machine. *IFAC-PapersOnLine*, *52*(30), 391–396. DOI: 10.1016/j.ifacol.2019.12.588

Harini, V. K., Meher, S. R., & Alex, Z. C. (2024). A novel refractive index based-fiber optic sensor for milk adulteration detection. *Optical Materials*, *154*, 115810. DOI: 10.1016/j.optmat.2024.115810

He, H., Sun, D. W., Pu, H., Chen, L., & Lin, L. (2019). Applications of Raman spectroscopic techniques for quality and safety evaluation of milk: A review of recent developments. *Critical Reviews in Food Science and Nutrition*, *59*(5), 770–793. DOI: 10.1080/10408398.2018.1528436 PMID: 30614242

He, Y., Bai, X., Xiao, Q., Liu, F., Zhou, L., & Zhang, C. (2021). Detection of adulteration in food based on nondestructive analysis techniques: A review. *Critical Reviews in Food Science and Nutrition*, *61*(14), 2351–2371. DOI: 10.1080/10408398.2020.1777526 PMID: 32543218

Hossen, R., Hossain, M. S., Ahmed, S., Rahman, M. M., & Sen, S. (2024). Illegal food chemicals sensing with photonic crystal fiber sensor in the terahertz spectrum. *Sensing and Bio-Sensing Research*, *44*, 100643. DOI: 10.1016/j.sbsr.2024.100643

Imam, M., & Nagpal, K. (2023). The electronic tongue: An advanced taste-sensing multichannel sensory tool with global selectivity for application in the pharmaceutical and food industry. *Pharmaceutical Development and Technology*, *28*(3-4), 318–332. DOI: 10.1080/10837450.2023.2194989 PMID: 36987792

Ionescu, A. D., Cîrîc, A. I., & Begea, M. (2023). A review of milk frauds and adulterations from a technological perspective. *Applied Sciences (Basel, Switzerland)*, *13*(17), 9821. DOI: 10.3390/app13179821

Ji, Z., Zhang, J., Deng, C., Guo, T., Han, R., Yang, Y., Zang, C., & Chen, Y. (2024). Identification of pasteurized mare milk and powder adulteration with bovine milk using quantitative proteomics and metabolomics approaches. *Food Chemistry: X*, *22*, 101265. DOI: 10.1016/j.fochx.2024.101265 PMID: 38468636

Jin, H., Dong, G. M., Wu, H. Y., Yang, Y. R., Huang, M. Y., Wang, M. Y., & Yang, R. J. (2023). Identification of adulterated milk based on auto-correlation spectra. *Spectrochimica Acta. Part A: Molecular and Biomolecular Spectroscopy*, *286*, 121987. DOI: 10.1016/j.saa.2022.121987 PMID: 36265304

Jitsuwan, P., Klinsophon, N., Thanakiatkrai, P., & Kitpipit, T. (2024). Simple closed-system combining direct PCR and fluorescence for all types of milk and dairy product authentication. *Journal of Food Composition and Analysis*, *130*, 106178. DOI: 10.1016/j.jfca.2024.106178

Kamboj, U., Kaushal, N., Mishra, S., & Munjal, N. (2020). Application of selective near infrared spectroscopy for qualitative and quantitative prediction of water adulteration in milk. *Materials Today: Proceedings*, *24*, 2449–2456. DOI: 10.1016/j.matpr.2020.03.775

Karunathilaka, S. R., Yakes, B. J., He, K., Brückner, L., & Mossoba, M. M. (2018). First use of handheld Raman spectroscopic devices and on-board chemometric analysis for the detection of milk powder adulteration. *Food Control*, *92*, 137–146. DOI: 10.1016/j.foodcont.2018.04.046

Kharajinezhadian, R., Chaichi, M. J., Nazari, O., Lakouraj, M. M., & Hasantabar, V. (2023). Fraud monitoring using a new disposable photoluminescence sensor in milk. *Microchemical Journal*, *189*, 108437. DOI: 10.1016/j.microc.2023.108437

Khosroshahi, M. E., Patel, Y., & Woll-Morison, V. (2022). Non-destructive assessment of milk quality using pulsed UV photoacoustic, fluorescence and near FTIR spectroscopy. *Laser Physics Letters*, *19*(7), 075602. DOI: 10.1088/1612-202X/ac6fc5

Kumar, R. R., Rahman, F., Bora, B., & Shameeh, M. (2023). Importance and nutritive value of animal proteins in human diet. In *Processing Technologies and Food Protein Digestion* (pp. 1–25). Academic Press. DOI: 10.1016/B978-0-323-95052-7.00007-8

Lal, P. P., Prakash, A. A., Chand, A. A., Prasad, K. A., Mehta, U., Assaf, M. H., Mani, F. S., & Mamun, K. A. (2022). IoT integrated fuzzy classification analysis for detecting adulterants in cow milk. *Sensing and Bio-Sensing Research*, *36*, 100486. DOI: 10.1016/j.sbsr.2022.100486

Lelis, C. A., Galvan, D., Tessaro, L., de Andrade, J. C., Mutz, Y. S., & Conte-Junior, C. A. (2022). Fluorescence spectroscopy in tandem with chemometric tools applied to milk quality control. *Journal of Food Composition and Analysis*, *109*, 104515. DOI: 10.1016/j.jfca.2022.104515

Mabood, F., Ali, L., Boque, R., Abbas, G., Jabeen, F., Haq, Q. M. I., Hussain, J., Hamaed, A. M., Naureen, Z., Al-Nabhani, M., Khan, M. Z., Khan, A., & Al-Harrasi, A. (2020). Robust Fourier transformed infrared spectroscopy coupled with multivariate methods for detection and quantification of urea adulteration in fresh milk samples. *Food Science & Nutrition*, *8*(10), 5249–5258. DOI: 10.1002/fsn3.987 PMID: 33133527

Mahanti, N. K., Shivashankar, S., Chhetri, K. B., Kumar, A., Rao, B. B., Aravind, J., & Swami, D. V. (2024). Enhancing Food Authentication through E-Nose and E-Tongue Technologies: Current Trends and Future Directions. *Trends in Food Science & Technology*, *150*, 104574. DOI: 10.1016/j.tifs.2024.104574

Majcher, M. A., Kaczmarek, A., Klensporf-Pawlik, D., Pikul, J., & Jeleń, H. H. (2015). SPME-MS-based electronic nose as a tool for determination of authenticity of PDO cheese, Oscypek. *Food Analytical Methods*, *8*(9), 2211–2217. DOI: 10.1007/s12161-015-0114-x

Mascini, M., & Tombelli, S. (2008). Biosensors for biomarkers in medical diagnostics. *Biomarkers*, *13*(7-8), 637–657. DOI: 10.1080/13547500802645905 PMID: 19061054

Moghaddam, H. N., Tamiji, Z., Lakeh, M. A., Khoshayand, M. R., & Mahmoodi, M. H. (2022). Multivariate analysis of food fraud: A review of NIR based instruments in tandem with chemometrics. *Journal of Food Composition and Analysis*, *107*, 104343. DOI: 10.1016/j.jfca.2021.104343

Mohammed, A., Potdar, V., & Quaddus, M. (2023). Exploring factors and impact of blockchain technology in the food supply chains: An exploratory study. *Foods*, *12*(10), 2052. DOI: 10.3390/foods12102052 PMID: 37238870

Mostafapour, S., Gharaghani, F. M., & Hemmateenejad, B. (2021). Converting electronic nose into opto-electronic nose by mixing MoS2 quantum dots with organic reagents: Application to recognition of aldehydes and ketones and determination of formaldehyde in milk. *Analytica Chimica Acta, 1170*, 338654. DOI: 10.1016/j.aca.2021.338654 PMID: 34090585

Moyer, D. C., DeVries, J. W., & Spink, J. (2017). The economics of a food fraud incident–Case studies and examples including Melamine in Wheat Gluten. *Food Control, 71*, 358–364. DOI: 10.1016/j.foodcont.2016.07.015

Nagraik, R., Sharma, A., Kumar, D., Chawla, P., & Kumar, A. P. (2021). Milk adulterant detection: Conventional and biosensor based approaches: A review. *Sensing and Bio-Sensing Research, 33*, 100433. DOI: 10.1016/j.sbsr.2021.100433

Nascimento, C. F., Santos, P. M., Pereira-Filho, E. R., & Rocha, F. R. (2017). Recent advances on determination of milk adulterants. *Food Chemistry, 221*, 1232–1244. DOI: 10.1016/j.foodchem.2016.11.034 PMID: 27979084

Othman, S., Mavani, N. R., Hussain, M. A., Abd Rahman, N., & Ali, J. M. (2023). Artificial intelligence-based techniques for adulteration and defect detections in food and agricultural industry: A review. *Journal of Agriculture and Food Research, 12*, 100590. DOI: 10.1016/j.jafr.2023.100590

Patel, A. S., Brahmbhatt, M. N., Bariya, A. R., Nayak, J. B., & Singh, V. K. (2023). Blockchain technology in food safety and traceability concern to livestock products. *Heliyon, 9*(6), e16526. DOI: 10.1016/j.heliyon.2023.e16526 PMID: 37265625

Patil, G. B., Wani, S. P., Bafna, P. S., Bagul, V. S., Kalaskar, M. G., & Mutha, R. E. (2024). Milk adulteration: From detection to health impact. *Food and Humanity*, 100339.

Pereira, T. D. S., dos Santos, D. M., Andre, R. D. S., & Correa, D. S. (2024). Zein/MnO2 Nanosheet Composites Integrated with a Smartphone for Colorimetric Sensors for On-Site Detection of Adulterants in Milk. *ACS Applied Nano Materials*.

Peris, M., & Escuder-Gilabert, L. (2016). Electronic noses and tongues to assess food authenticity and adulteration. *Trends in Food Science & Technology, 58*, 40–54. DOI: 10.1016/j.tifs.2016.10.014

Poghossian, A., Geissler, H., & Schöning, M. J. (2019). Rapid methods and sensors for milk quality monitoring and spoilage detection. *Biosensors & Bioelectronics, 140*, 111272. DOI: 10.1016/j.bios.2019.04.040 PMID: 31170654

Qin, C., Liu, L., Wang, Y., Leng, T., Zhu, M., Gan, B., Xie, J., Yu, Q., & Chen, Y. (2022). Advancement of omics techniques for chemical profile analysis and authentication of milk. *Trends in Food Science & Technology, 127*, 114–128. DOI: 10.1016/j.tifs.2022.06.001

Rahman, A., Habib, M. R., Ali, M. Y., Islam, M. A., & Rashid, M. H. U. (2017). Physico-chemical analysis and detection of adulteration in raw milk collected from Goalas of different places of sadar upazila in Mymensingh district. *Research in Agriculture Livestock and Fisheries, 4*(2), 99–106. DOI: 10.3329/ralf.v4i2.33721

Rebechi, S. R., Vélez, M. A., Vaira, S., & Perotti, M. C. (2016). Adulteration of Argentinean milk fats with animal fats: Detection by fatty acids analysis and multivariate regression techniques. *Food Chemistry, 192*, 1025–1032. DOI: 10.1016/j.foodchem.2015.07.107 PMID: 26304443

Sarkar, D., Ranjan, R., Marwaha, S., Sahoo, A., & Naha, S. (2024). Multiplex PCR for detection of camel milk adulteration with cattle and goat milk. *International Dairy Journal, 154*, 105922. DOI: 10.1016/j.idairyj.2024.105922

Sen, S., Dundar, Z., Uncu, O., & Ozen, B. (2021). Potential of Fourier-transform infrared spectroscopy in adulteration detection and quality assessment in buffalo and goat milks. *Microchemical Journal, 166*, 106207. DOI: 10.1016/j.microc.2021.106207

Sharma, A., Batra, N., Garg, A., & Saxena, A. (2017). Food adulteration: A review. *International Journal for Research in Applied Science and Engineering Technology, 5*(11), 686–689. DOI: 10.22214/ijraset.2017.3129

Sharma, R., Rajput, Y. S., & Barui, A. K. (2012). *Detection of adulterants in milk: laboratory manual*. NDRI.

Shawky, E., Nahar, L., Nassief, S. M., Sarkcr, S. D., & Ibrahim, R. S. (2024). Dairy products authentication with biomarkers: A comprehensive critical review. *Trends in Food Science & Technology, 147*, 104445. DOI: 10.1016/j.tifs.2024.104445

Singh, A., Chandra, G., Aggarwal, A., & Kumar, P. (2012). Adulteration detection in milk. *Res News U, 5*, 52–55.

Spink, J. W. (2019). *Food fraud prevention: introduction, implementation, and management*. Springer Nature. DOI: 10.1007/978-1-4939-9621-6

Suman, M., Cavanna, D., Sammarco, G., Lambertini, F., & Loffi, C. (2021). Fighting food frauds exploiting chromatography-mass spectrometry technologies: Scenario comparison between solutions in scientific literature and real approaches in place in industrial facilities. *Trends in Analytical Chemistry, 142*, 116305. DOI: 10.1016/j.trac.2021.116305

Tehrani, T., Pont, L., & Benavente, F. (2024). Rapid detection and quantification of milk adulteration using MALDI-MS protein profiling and multivariate calibration. *Journal of Food Composition and Analysis, 130*, 106147. DOI: 10.1016/j.jfca.2024.106147

Tian, H., Chen, S., Li, D., Lou, X., Chen, C., & Yu, H. (2022). Simultaneous detection for adulterations of maltodextrin, sodium carbonate, and whey in raw milk using Raman spectroscopy and chemometrics. *Journal of Dairy Science, 105*(9), 7242–7252. DOI: 10.3168/jds.2021-21082 PMID: 35863924

Tian, H., Xiong, J., Chen, S., Yu, H., Chen, C., Huang, J., Yuan, H., & Lou, X. (2023). Rapid identification of adulteration in raw bovine milk with soymilk by electronic nose and headspace-gas chromatography ion-mobility spectrometry. *Food Chemistry: X, 18*, 100696. DOI: 10.1016/j.fochx.2023.100696 PMID: 37187488

Titova, T., & Nachev, V. (2020). Electronic tongue" in the food industry. *Food Science and Applied Biotechnology, 3*(1), 71–76. DOI: 10.30721/fsab2020.v3.i1.74

Tsenkova, R., & Muncan, J. (2022). Milk quality management: state of the art and contemporary requirements. In *Aquaphotomics for Bio-diagnostics in Dairy*. Springer. DOI: 10.1007/978-981-16-7114-2_1

Valand, R., Tanna, S., Lawson, G., & Bengtström, L. (2020). A review of Fourier Transform Infrared (FTIR) spectroscopy used in food adulteration and authenticity investigations. *Food Additives & Contaminants. Part A, Chemistry, Analysis, Control, Exposure & Risk Assessment, 37*(1), 19–38. DOI: 10.1080/19440049.2019.1675909 PMID: 31613710

Vasconcelos, H., Matias, A., Jorge, P., Saraiva, C., Mendes, J., Araújo, J., & Coelho, L. C. (2021). Optical biosensor for the detection of hydrogen peroxide in milk. *Chemistry Proceedings, 5*(1), 55.

Vern, P., Panghal, A., Mor, R. S., & Kamble, S. S. (2024). Blockchain technology in the agri-food supply chain: a systematic literature review of opportunities and challenges. *Management Review Quarterly*, 1-33.

Vinciguerra, L. L., Marcelo, M. C., Motta, T. M., Meneghini, L. Z., Bergold, A. M., & Ferrão, M. F. (2019). Chemometric tools and FTIR-ATR spectroscopy applied in milk adulterated with cheese whey. *Quimica Nova, 42*(3), 249–254. DOI: 10.21577/0100-4042.20170337

Vinothkanna, A., Dar, O. I., Liu, Z., & Jia, A. Q. (2024). Advanced detection tools in food fraud: A systematic review for holistic and rational detection method based on research and patents. *Food Chemistry*, *446*, 138893. DOI: 10.1016/j.foodchem.2024.138893 PMID: 38432137

Visciano, P., & Schirone, M. (2021). Food frauds: Global incidents and misleading situations. *Trends in Food Science & Technology*, *114*, 424–442. DOI: 10.1016/j.tifs.2021.06.010

Whyte, S. T., Omoyiola, B. O., & Okoni, B. (2022). Use of Blockchain Technology in Data Integrity Assurance. *Available atSSRN* 4043164. DOI: 10.2139/ssrn.4043164

Windarsih, A., Arifah, M. F., Utami, I. D., Suratno, , & Rohman, A. (2024). Detection of goat milk adulteration in horse milk using LC-HRMS-based non-targeted metabolomics and chemometrics. *Chemicke Zvesti*, *78*(2), 809–821. DOI: 10.1007/s11696-023-03123-5

Yaman, H. (2020, August). A rapid method for detection adulteration in goat milk by using vibrational spectroscopy in combination with chemometric method s. *Journal of Food Science and Technology*, *57*(8), 3091–3098. DOI: 10.1007/s13197-020-04342-4 PMID: 32624611

Yang, J., Zheng, N., Soyeurt, H., Yang, Y., & Wang, J. (2019). Detection of plant protein in adulterated milk using nontargeted nano-high-performance liquid chromatography–tandem mass spectroscopy combined with principal component analysis. *Food Science & Nutrition*, *7*(1), 56–64. DOI: 10.1002/fsn3.791 PMID: 30680159

Yashaswini, N., Suraj, L., Nayak, S., Khosla, A., & Manjunatha, C. (2022). Construction, Working, and Applications of E-Tongue: A Versatile Tool for All Tastes? *ECS Transactions*, *107*(1), 20193–20211. DOI: 10.1149/10701.20193ecst

Zhao, V. X. T., Wong, T. I., Zheng, X. T., Tan, Y. N., & Zhou, X. (2020). Colorimetric biosensors for point-of-care virus detections. *Materials Science for Energy Technologies*, *3*, 237–249. DOI: 10.1016/j.mset.2019.10.002 PMID: 33604529

Zhu, C., Fu, X., Zhang, J., Qin, K., & Wu, C. (2022). Review of portable near infrared spectrometers: Current status and new techniques. *Journal of Near Infrared Spectroscopy*, *30*(2), 51–66. DOI: 10.1177/09670335211030617

Chapter 3
Role of Extra Cellular Polysaccharides in Dairy Products

Abdullah Ahmed Butt
https://orcid.org/0009-0008-1811-9671
Government College University, Faisalabad, Pakistan

Zahra Ahmed
King Edward Medical University, Lahore, Pakistan

ABSTRACT

Extracellular polysaccharides play a critical role in augmenting the quality and functionality of dairy products. This chapter provides an extensive exploration of EPS, highlighting the types, functions, and applications in different dairy matrices. EPS in dairy products is mainly produced by lactic acid bacteria. It is essential to note that EPS can be linked to the texture, viscosity, and gelation properties of dairy products. These characteristics are decisive for thus organoleptic properties and consumers' desirability for food products, such as, e.g., yogurt, cheese, and ice cream. In addition to technological aspects, EPS is associated with numerous health benefits, such as probiotic effects, immune modulation, and potential prebiotic action. However, this chapter also focuses on the technological issues of both processing and utilizing EPS for dairy products. These elements should be further researched because they can help optimize use and facilitate the process of production of new, high quality food products that meet consumers' and regulatory agencies' requirements

DOI: 10.4018/979-8-3693-9949-1.ch003

INTRODUCTION

Overview of Polysaccharides in Dairy Products

Complex carbohydrates and polysaccharides are made up of units of monosaccharide repeating units attached to glycosidic bonds. As thickeners, stabilizers, emulsifiers, and water binding agents they have multiple roles in food systems. Polysaccharides from lactic acid bacteria (LAB) play an important role in the texture, stability, and overall sensory quality of dairy products. During fermentation processes, the ability to synthesize extracellular polysaccharides (EPS) from naturally occurring sources allows for their integration into the dairy matrix these EPS are highly desirable for clean label and functional food applications (Prete et al., 2021). In dairy processing, polysaccharides can originate from two primary sources, in situ production during fermentation by LAB, or using exogenous plant or microbial polysaccharides such as xanthan gum and guar gum. The interest in the latter part is because of the fact that it is also responsible for the dairy product's natural and organic appeal. The production and properties of microbial EPS are, however, determined by factors such as bacterial strain specificity, nutrient availability, and fermentation conditions. This gives it considerable adaptability for utilization in various dairy applications. Polysaccharides play extensive functional roles in dairy products. Increasing viscosity and creaminess in products like yogurt gel and they stabilize emulsions avoiding defects such as syneresis (Tiwari et al., 2021). The function of these properties not only enhances product quality but also extends shelf life. EPS act as dietary fibers due to prebiotic potential, and are good for gut health by promoting the growth of good microbes (Muninathan et al., 2022). Immunomodulatory as well as metabolic health benefits of some polysaccharides have contributed to their importance in functional dairy foods development. The EPS production rate is usually calculated by.

$$EPS/\ dt = Y_{EPS} * \mu * X$$

Where EPS/ dt is the rate of EPS measured in g/L/h

Y_{EPS} is the output coefficient of EPS, g EPS obtained per gram of product under test

μ is the growth rate measured in h^{-1}

X is biomass concentration usually measured in g/L

From yogurt, and cheese to ice cream, polysaccharides are used. EPS in yogurt improves mouthfeel and yogurt stability, whereas EPS in cheese increases water retention, texture, and curd formation. Freezing of desserts stabilized by polysaccharides reduces ice crystal formation and improves freeze-thaw stability. Further, these contributions are important to fulfill consumer needs for premium quality, health-promoting, and sustainable dairy products. In this chapter, the various functionalities of polysaccharides as sources, health benefits, and applications in dairy products as well as functionality are also discussed. Current advancements and barriers to advance towards optimal polysaccharide utilization in innovative and sustainable dairy operations are discussed.

Importance of Extracellular Polysaccharides (EPS)

In the dairy processing industry, EPS has been widely documented for its versatile roles in the technological and nutritional properties of products. They are important because they can offer benefits for the structural and sensory quality of dairy products, increase production efficiencies, and supply health benefits, and are indispensable to modern dairy innovations. From a functional viewpoint, EPS is used to alter the rheological characteristics of dairy products. Apart from the fundamental importance of consumer satisfaction, viscosity enhancing, creaminess enhancing, and improving mouthfeel are also achieved. The interaction of milk proteins and EPS forms stable gel networks in fermented yogurt contributing to secondary viscosity, preventing syneresis, and strengthening texture stability over time. In cheese, EPS helps retain water, contributes to curd texture, and enhances processing cheese melting (Bihola et al., 2024). They also act as natural stabilizers furthering the demand for clean label and minimally processed foods, while their ability to act as natural stabilizers also reduces the need for synthetic additives.

EPS has also technological applications. They are classified as dietary fibers that resist digestion in the human gastrointestinal tract and are substrates to beneficial gut microbiota. The gut health, nutrient absorption, and overall health support its prebiotic potential. Like many other EPS, some also have immunomodulatory properties that help to regulate immune response and reduce inflammation. Additionally, this can alleviate lactose intolerance when their inclusion in lactose-containing dairy products results in their growth as probiotics to assist with lactose digestion. EPS not only improves product quality but facilitates sustainable and innovative food production. It enhances the development of organic and functional dairy products as their natural synthesis occurs during fermentation processes (Prete et al., 2021). Research and industry professionals pay close attention to EPS as consumer preferences continue to change, and as no solutions currently exist to completely address this challenge. Optimization, and their potential health benefits remain a focus in dairy formulation. Their multifunctional nature highlights the major role that they play in improving dairy technology and meeting technological and nutritional demands.

TYPES OF EXTRACELLULAR POLYSACCHARIDES

Classification of EPS

Classification Based on Chemical Composition

Homogeneous polysaccharides (HoPS) and heterogeneous polysaccharides (HePS) are the two broad categories of EPS, as the monosaccharide composition of EPS is heterogeneous. A homopolysaccharide is composed of only one monosaccharide unit. Some examples of dextran, produced by Leuconostoc species, and levan, synthesized by Lactobacillus species are popular. These polysaccharides are high molecular weight with water-binding properties and are used as a thickening or stabilizing agent. Heteropolysaccharides, on the other hand, contain more than one type of monosaccharide like glucose, galactose, and rhamnose. These have normally more complex structures and a much wider range of functional properties. LAB helps in dairy fermentations and they play a major role in the textural and stabilizing aspects of such fermented products as yogurt, and cheese (Sharma et al., 2023).

Classification Based on Structural Properties

EPS can be classified into linear and branched polysaccharides according to their molecular structure. The linear EPS has a simple unbranched chain of monosaccharide units. Gelation and viscosity enhancement in food systems are often attributable to these structures. The side chains branch from the main polymer backbone in EPS species. The resulting branching endows these dairy products with such rheological properties as increased elasticity and better emulsion stability.

Based on Charge Classifier

Another important parameter for classification is the charge of the polysaccharide molecule. Neutral EPS have no charged groups and are usually hydrophobic. Most of these are used to modify texture and to improve water-holding capacity. On the other hand, charged EPS have anionic or cationic groups, e.g., phosphate, sulfate, or carboxyl groups. To be able to interact with proteins and minerals in the dairy matrix, these charges allow them to form stable structures as well as influence product rheology (Yang et al., 2022).

Classification According to Mode of Production

The microbial cell surface-associated EPS is distinguished as capsular polysaccharides (CPS) or slime polysaccharides. A protective layer is formed by tightly bound Capsular Polysaccharides (CPS) to the cell surface. Indirectly stabilizing microbial culture, CPS contributes to biofilm formation and cell aggregation, impacting positively dairy fermentation. Slime Polysaccharides are loosely attached or secreted into the surrounding medium. These polysaccharides affect directly the texture and viscosity of fermented dairy products (Angelin & Kavitha, 2020).

Figure 1. Classification of EPS

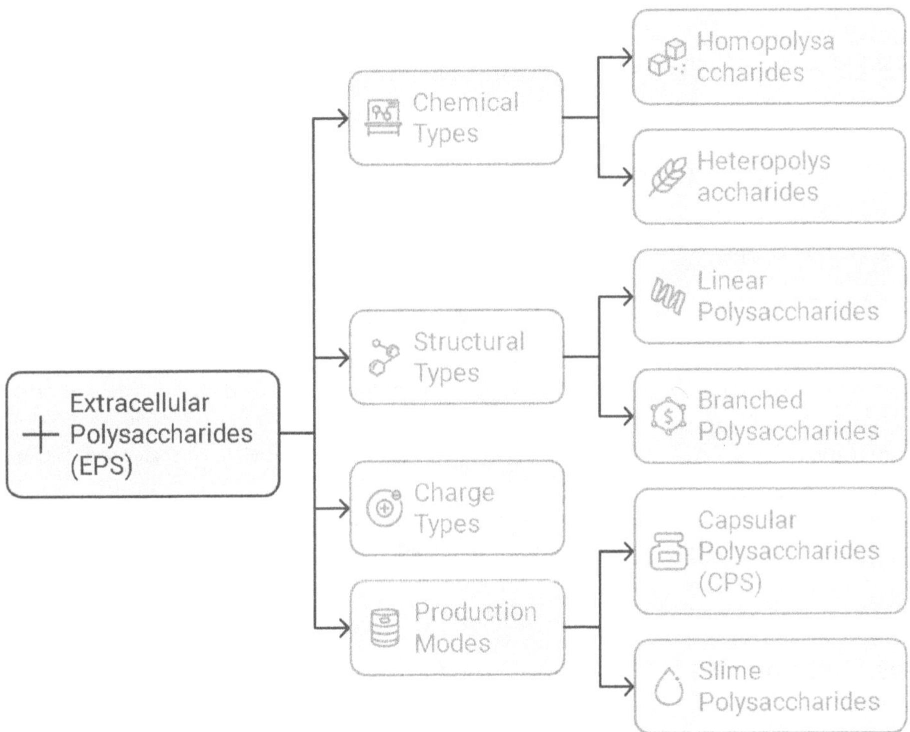

Sources of EPS in Dairy Fermentations

Lactic Acid Bacteria (LAB) as Primary Producers

EPS in milk is predominantly produced in dairy fermentations by LAB. Common EPS-producing LAB strains include species of Lactobacillus, Lactococcus, Leuconostoc, and *Streptococcus thermophilus (S. thermophilus)*. EPS is synthesized by these bacteria during their normal metabolism, normally in response to stress conditions, particularly when restricted for nutrients, when temperature or pH changes. Each species possesses different capabilities for producing the EPS and the type and amounts of these are strain specific. As example, *Lactobacillus delbrueckii subsp. Bulgaricus (L. delbrueckii)* and *S. thermophilus* are commonly used in yogurt fermentation since their synergistic EPS production increases the product's viscosity and mouthfeel (Xu et al., 2021).

Table 1. Common LABs producing EPS and their applications

LAB strain	EPS type	Functional properties	Dairy Applications
Lactobacillus rhamnosus (L. rhamnosus)	Rhamnose-based polysaccharides	Improves viscosity and mouthfeel	Yogurt, fermented milk products
L. delbrueckii	Glucan-based polysaccharides	Enhances water retention, reduces syneresis	Cheese, fermented dairy products
Bifidobacterium lactis (B. lactis)	Fructan-based polysaccharides	Modulates gut health, acts as a prebiotic	Probiotic dairy beverages, yogurt
S. thermophilus	Galactan-based polysaccharides	Enhances texture and stability	Yogurt, soft cheeses
Lactobacillus casei (L. casei)	Glucans and galactans	Enhances probiotic properties, improves texture	Cheese, yogurt, ice cream

Milk Components Interaction

The composition of the milk substrate affects EPS production in dairy fermentations. LABs are mainly supplied with milk carbohydrates, and lactose, which are used as the main carbon source for EPS biosynthesis. EPS production is also modulated by the availability of minerals, proteins, and other micronutrients in milk. For example, usages of dairy-based EPS, for instance, could be affected by calcium ions' influences on the gelation and rheological properties of the EPS or the functional role of these biopolymers.

Cocultures and Mixed Starter Cultures, Role

Mixed or coculture of LAB is used in many dairy fermentations to improve EPS production. By using multiple strains, they can improve EPS quantity and synthesis diversity by the complementarity of the metabolic pathways of different species. For instance, EPS production by *S. thermophilus* increases gel strength and reduces syneresis significantly in yogurt, when in combination with *L. bulgaricus (Arab et al., 2023)*. The importance of selecting compatible strains for desirable EPS-related characteristics is shown in this in the following sections.

Non-LAB EPS Producers

Though LAB is the major EPS producer in dairy, other microorganisms, including Bifidobacterium and Propionibacterium, also produce EPS. EPS from these bacteria are often used in probiotic formulations to provide health benefits such as prebiotic effects and immune modulation but also improve the textural properties of the product by acting as a gelling agent (Duboux et al., 2021). In addition, the addition of such microorganisms enlarges the functional and nutritional range of fermented dairy products.

Strain Optimization and Genetic Engineering.

Because advancements in genetic engineering have proved that it is possible to develop LAB strains showing increases in synthesis of EPS. Researchers have enabled an increase in the yield and functional diversity of EPS by modulating EPS biosynthetic metabolic pathways. Specifically, these engineered strains are useful in dairy fermentations where textural or health-promoting properties are desired, but not required.

Structural Characteristics of EPS

Monosaccharide Composition

Repeating monosaccharide units such as glucose, galactose, rhamnose, mannose, fructose, and occasionally uronic acids are the primary monomers from which the monomers of EPS are composed. The functional properties of the polymer are defined by a specific combination of the monosaccharides. For example, the anionic properties of the EPS associated with the presence of uronic acids provide an opportunity for binding to proteins and minerals used for the fabrication of stable gel structures in fermented dairy products.

Heteropolysaccharides and Homopolysaccharides

Homopolysaccharides (HoPS) and heteropolysaccharides (HePS) constitute an EPS depending on the composition of constituent monosaccharides. Homopolysaccharides are those of a single type of monosaccharide. Dextran (glucose polymers) and levan (fructose polymers), for example, have high molecular weight, high viscosity, and gelling capability. Heteropolysaccharides comprise two or more monosaccharides in repeating units but with the property to be different from each other. Dairy fermentations and resulting HePS are characterized by more complex branched structures and often have unique functional properties associated with enhanced texture and emulsion stability.

Molecular Weight

Widely dispersed molecular weight (MW) of EPS, from a few kilodaltons to several megadaltons, determines EPS functionality. EPS with high molecular weights tend to form viscous solutions and act to gelate, while EPS with low molecular weights are more soluble and can be used as emulsifiers or stabilizers (Akan & Oner, 2022). The molecular weight of synthesized materials is dictated by the enzymatic products of the producing microorganisms and can be controlled either through fermentation conditions or through genetic engineering.

Linkage Types

EPS is glycosidically linked monosaccharides by glycosidic bonds which can be α or β. The polymer's solubility, rigidity, and digestibility depend on the type of linkage. For example, EPS composed of β linked polysaccharides are more resistant than the other to enzymatic degradation and therefore play a role in the prebiotic properties of certain EPS. The diversity of EPS organization is further increased, the position of the linkage, for instance, being 1→3, 1→4, or 1→6 (Zhou et al., 2019).

Degree of Branching

The degree of branching of the EPS is important for their rheological behavior. EPS, with multiple side chains, exhibits high water binding capacity and improves viscosity so that it can stabilize dairy products. The great advantage gained from linear EPS is its potential to align and interact with other polymer chains and for forming gels.

Charge and Functional Groups

EPS can be either neutral or charged, but only when they have functional groups, such as carboxyl or sulfate groups. Charged EPS interacts in unique ways with proteins and other components of the dairy to form stable complexes responsible for the textural and sensory properties of dairy products such as yogurt and cheese. Anionic EPS can for example bind positively charged calcium ions which affect gelation and curding.

FUNCTIONAL ROLES OF EPS IN DAIRY PRODUCTS

Textural Enhancement

Milk Proteins Interactions

The interaction of EPS with milk proteins is one of the principal mechanisms by which EPS contributes to texture. Fermented products like yogurt are used to produce a protein network reinforced by a gel matrix formed in situ by EPS produced by LAB. As these interactions increase the gel's strength, and elasticity, it becomes smoother and creamier. But as we raise the gel's strength and elasticity with these interactions, this becomes smoother and creamier.

Tailoring Textural Properties

This structural diversity variation in molecular weight, monosaccharide composition, and branching patterns is what allows variation in textural properties independent of the 'perfect' polymer chain to meet requirements in specific applications. For one example, linear EPS is better used to form firm gels, while highly branched EPS with high molecular weight is more effective in creating thick and creamy textures (Kumar et al., 2022). The same characteristics can be finetuned by selecting certain EPS-producing strains or optimizing fermentation conditions.

Figure 2. Roles played by EPS in textural enhancement of dairy products

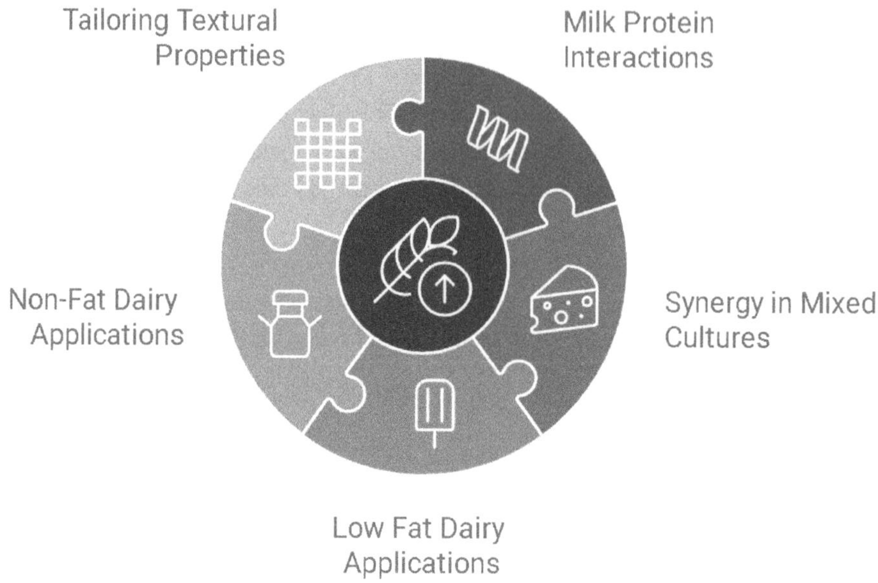

Viscosity and Gelation Properties

Dairy Matrix Viscosity Enhancement

Fermented dairy products like yogurt, sour cream, and buttermilk have high EPS viscosity contribution. The viscosity-enhancing properties of EPS result from their ability to form entangled networks in the aqueous phase of milk, thereby increasing the resistance to flow. The high molecular weight and highly branched EPS result in more robust networks, leading to thicker textural behaviors (richer) with a creamier texture. For example, viscosity improvers such as dextran from *Leuconostoc mesenteroides (L. mesenteroides)* and certain HePS from *S. thermophilus* are remarkably effective. EPS not only plays a role in enhancing sensory properties but also in stabilizing the product against phase separation, both major causes for product rejection (Berthold-Pluta et al., 2019). EPS binds water by forming a viscous matrix that reduces syneresis, or whey separating, often a problem with fermented dairy products. It is a stabilizing effect that makes texture quality and visual quality improve and consistent texture without disturbance.

Gelation Properties and Structural Integrity

In addition to viscosity increase, EPS also is important for dairy product gelation. During fermentation, EPS becomes associated with casein micelles and other milk proteins to reinforce the gel network formed upon acid-induced coagulation. The interaction improves the stability (elasticity, firmness, and cohesiveness) of the gel making the product more stable and more appealing. For example, EPS from *S. thermophilus* and *L. bulgaricus* are essential contributors to the manufacture of a smooth and speckled gel with little syneresis. Particularly useful in low-fat and nonfat dairy products where the lack of fat can produce weak or brittle gels, the gelation properties of EPS are very desirable (Basiri, 2021). Due to the loss of fat, the matrix is strengthened by EPS and enables a promising texture and mouthfeel resilience of the product. Meeting consumers' preferences for healthy but indulgent dairy options is dependent upon this functionality.

Rheological Behavior

EPS rheological behavior including their tendency to show shear-thinning or pseudoplastic properties improves their use in dairy products. For products such as yogurt and drinkable dairy formulations, where the viscosity falls under applied shear stress, is desirable to promote ease of handling and consumption with a thick texture. The selection of strains of LAB that produce EPS with different structural features, such as different monosaccharide compositions or linkage types, allows tailoring of the specific rheological properties of EPS.

Synergistic Effects in Mixed Cultures

Both viscosity and gelation can be improved by the EPS production of different LAB strains in mixed culture fermentations. For example, the EPS produced by *S. thermophilus* and *L. bulgaricus*, work in a complementary way to produce a stronger, and more elastic gel network. These synergistic effects permit cost-effective facilitation of textural and rheological properties (Atik et al., 2024).

Figure 3. Viscosity and gelation properties of EPS

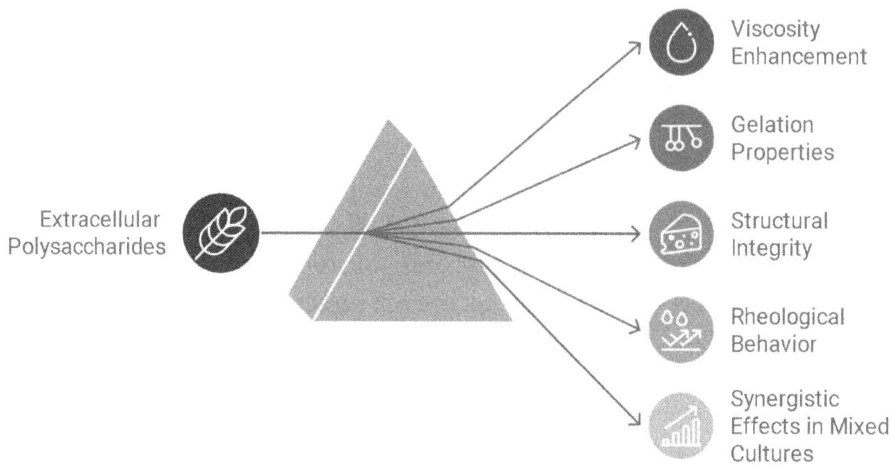

Water Holding Capacity and Syneresis Control

Water Holding Capacity (WHC)

Since EPS can form hydrophilic networks that greatly hold water, they are extremely efficient in increasing the WHC of dairy products. The molecular structure of most of these is primarily due to their presence of hydrophilic functional groups like hydroxyl and carboxyl groups (Zhang et al., 2024). The product interacts with these groups that in turn interact with water molecules and make the product water-retentive. EPS produced by LAB, including Lactobacillus and *S. thermophilus*, play a major role in improving the WHC of yogurt and other fermented milk products in dairy fermentations. This is important since EPS holds water well and water is critical in products with lower fat content, water retention makes up for the lack of fat and contributes towards a creamy and smooth texture. High WHC also helps make the product more attractive to consumers due to better mouthfeel (Xu et al., 2024). Researchers have calculated a formula for the WHC of EPS.

WHC= Weight of water retained/ Total weight of sample * 100

Dairy Products Syneresis Control

In some situations, the release of whey from the gel matrix is unwanted, because it may result in loss of product quality, texture, and consumer acceptability. EPS plays an important role in controlling syneresis by forming a dense and stable three dimensional network that traps and retards water during storage and handling. Whey separation tendency in fermented dairy products is reduced, as whey separation is decreased by the water binding properties of higher molecular weight EPS with branched structures (Xu et al., 2024). The protein matrix is more structurally stable and is synergistically less prone to syneresis because of EPS increases structural integrity of the protein matrix and provides a more cohesive gel resistant to the physical forces that normally lead to syneresis. The EPS produced by Lactobacillus strains in yogurt improves product yield and results in a smoother texture by reducing whey separation, for example (Wang et al., 2022). Usually following formula is used to calculate the syneresis rate in dairy products.

$$\text{Syneresis} = V_{initial} - V_{final} / V_{initial} * 100$$

$V_{initial}$ is the initial volume of the gel or yogurt

V_{final} is the volume after syneresis (after liquid expulsion).

EPS Action in Water Retention Mechanism

EPS forms hydrophilic networks that interact with water and milk proteins and thereby enhance water retention and control syneresis. These networks trap water in a gel-like structure that does not force the water out from the gel when the gel undergoes physical stress (Prasad & Purohit, 2023). The viscosity and elasticity of EPS are critical factors in this process; both must be highly viscous and elastic sufficient for their EPS to form stronger, more stable gels that are stable under varying storage conditions. In addition, EPS that exhibit a higher degree of branching has a greater ability to hold water. Linear EPS generally does not perform as well as branched EPS because they typically have a larger surface area and more complicated polysaccharide chain networks. Branched EPS has this structural characteristic and is very valuable in preventing syneresis products such as yogurt, whose quality attributes depend more on texture and consistency.

Mixed Starter Cultures: Synergistic Effects

The use of mixed starter cultures of several different LAB strains in dairy fermentations allows the production of mixed EPS with complementary properties, leading to additional water retention and syneresis control. For example, EPS produced by *S. thermophilus* and *L. bulgaricus* in combination as in yogurt fermentation or similar matrix fermentation aids in enhanced water holding capacity and retardation of whey separation. The sum of these strains contributes to a smoother, more stable product with reduced syneresis.

Low Fat and Non-Fat Dairy Products

In low-fat and nonfat dairy products where increased syneresis and decreased texture can result from a slight reduction in fat, EPS is most beneficial. EPS helps provide a creamy and stable texture while reducing fat levels by improving water retention and preventing the instance of whey separation. The development of healthier dairy alternatives requires robust mouthfeel and quality but meeting consumers' expectations is highly reliant on this functionality (Appiani et al., 2023).

Impact on Mouthfeel and Sensory Attributes

Creaminess and Smoothness

The most noticeable effect of EPS in dairy products is the ability of EPS to impart a creamy smooth mouthfeel, particularly of interest in yogurt, sour cream, and dairy beverages. The fact that a particular EPS may have a high molecular weight, or have a complex branched structure, allows the molecular structure of the EPS to create stable, hydrated networks, that mimic the texture of fat (Lu et al., 2019). The outcome is a creamy velvety mouthfeel that complements the eating experience in itself. EPS in low-fat or fat-free dairy products replaces fat, thus improving their smoothness and creaminess without reducing the indulgence levels of these products.

Viscosity and Sensory Perception

Changes in mouthfeel are directly related to the viscosity-enhancing properties of EPS. A thicker, more cohesive texture is perceived as being more substantial and satisfying in the mouth. Dairy products such as yogurt can have their body and consistency made more pleasing by the addition of EPS, dextran, levan, and other heteropolysaccharides produced by *S. thermophilus* and Lactobacillus spp (Pourjafar et al., 2023). Because viscosity effects are particularly pronounced in yogurt, where increasing viscosity is frequently paralleled by a richer mouthfeel, these effects merit further investigation. The EPS shear thinning behavior, which slows viscosity under stress, contributes to a smooth texture during consumption, making it easy to swallow while keeping the overall richness when at rest.

Help Reduce Off Flavors, and Improve Taste

Moreover, EPS can affect the sensory profile of dairy products by allowing some off flavors or unpleasant tastes to mask into the product during fermentation (Tribby & Teter, 2023). EPS binding capacity for water and milk proteins to form stable gel-like structures can modulate taste perception by decreasing the bitterness or acidity perception often present in low-fat or fermented dairy products. EPS adds to this better texture masking to deliver a better balanced, palatable flavor profile and overall consumer sensory experience.

Gel Strength and Mouthfeel

Another critical factor influencing mouthfeel is the ability of EPS to make the gel network stronger in dairy products. For example, in yogurt, egg white extract can impart firmness and elasticity to the product thanks to the combination of the casein proteins' gel network and EPS network (Leal et al., 2024; Yang et al., 2024). In set-type yogurts, the consumers expect some degree of rigidity in the mouthfeel and the traditional firm gel form of a yogurt has often been associated with this form of mouthfeel. It enhances the structural integrity of the product and increases its appealing sensory and texture attributes by improving the interaction between EPS and milk proteins in forming a more robust gel matrix.

Sensory Impact in Mixed Cultures

EPS produced in mixed culture fermentations enhances sensory properties when different LAB strains are used in combination. *L. bulgaricus* and *S. thermophilus* produce EPS with different structural characteristics, whose synergism increases the texture and mouthfeel of yogurt. These EPS have an additive effect of smoothing, creaming, and uniformity of the product, enhancing the sensory attributes and, importantly, ours to a more balanced and enjoyable eating experience.

Influence on Consumer Preferences.

Dairy product mouthfeel, largely determined by EPS, is among the most important factors influencing consumer preference. Often, it's sensory tests that reveal that consumers prefer products with higher viscosity and creamier, smoother textures over thinner, watery products. In addition, the texture characteristics of yogurt and kefir products with a high content of EPS are higher rated for creaminess and mouthfeel satisfaction (Vieira et al.). Due to this, using EPS in dairy product formulations enables manufacturers to meet the changing requirements of palatable, and health-conscious food options from changing consumer demands.

Sensory Attributes Tailoring for Particular Products

EPS can create the desired sensory impact for the dairy product as desired. Manufacturers have selected EPS producing LAB strains which produce polysaccharides with various molecular weights or polysaccharide compositions to finetune mouthfeel and other sensory properties. To take a simple example, EPS with lower molecular weight may be employed in drinkable yogurts where a smooth, easy-to-consume product is desired without sacrificing mouthfeel. For set-type yogurts, however, preferred EPS to promote gel formation are of higher molecular weight and lead to a thicker more structured product (X. Yin et al., 2024).

Table 2. Effect of different EPS on sensory properties in different dairy products

Dairy Product:	**EPS type:**	**Sensory properties:**
Yogurt	Glucans, galactans	Improves smoothness, mouthfeel, and consistency
Soft cheese	Polysaccharides from *L. delbrueckii*	Creamier texture and improved flavor perception
Ice cream	Exopolysaccharides from *L. rhamnosus*	Creamy, velvety mouthfeel, better texture retention
Fermented milk	Fructans from *B. lactis*	Mild acidity, enhanced smoothness, and body

APPLICATIONS OF EPS IN SPECIFIC DAIRY PRODUCTS

Yogurt and Fermented Milk

Texture and Viscosity Contribution

EPS has one of the most important roles on texture and viscosity in yogurt and fermented milk. LAB strains, such as *S. thermophilus*, *L. bulgaricus*, and *Lactococcus lactis (L. lactis)*, can produce large amounts of EPS which can enhance the viscosity of the final product due to the formation of hydrophilic networks trapping water in the gel matrix. The consequence is a thicker, creamier texture that is exactly what people like in yogurt and other fermented milk beverages. Viscosity enhancement is dependent on the molecular weight and structure of the EPS. One example is that high molecular weight, branched polysaccharides usually give stronger and more cohesive gels, which results due to interaction of protein and polysaccharides (Goh et al., 2020). The higher viscosity not only improves the mouthfeel of yogurt by increasing its mass, but it also reduces syneresis (dropping liquid out of the gel), a typical problem in fermented dairy products. EPS prevents the expulsion of whey (in aqueous solution) by binding water and thus forces the textures of the product to remain as desired for its entire shelf life (Wang et al., 2023).

Syneresis and Gel Formation Control

Lactic acid bacteria lower the milk's pH during fermentation to cause the casein proteins to form a gel-like casein structure. LAB-produced EPS interacts with casein micelles to accelerate the formation of firm, cohesive gel networks. Because yogurt texture and structural integrity are dependent on this gelation, it is critical. The improved properties of gel formation lead to a product with a hardness and elasticity suitable for consumer preference. Furthermore, EPS involvement in controlling syneresis in yogurt (the undesirable phenomenon where whey is separated from the gel matrix) has also been researched. EPS enables the product to hold more water, thereby reducing the risk of whey separation into the product, leading to a product that is more stable especially in products like cheese and yogurt (Surber et al., 2021). Excessive syneresis can be deleterious for product texture and consumer appeal in yogurt, particularly because this phenomenon is difficult to avoid.

Mouthfeel and Sensory Attributes

EPS is very important for influencing the mouthfeel of yogurt and fermented milk products. EPS provides a creamy, smooth texture making the sensory experience better. EPS in low-fat or fat-free dairy products fill a need by improving creaminess without adding calories, where a lack of fat often leads to a less satisfying, thinner product (Kumar et al., 2022). This is particularly useful in the development of healthier yogurt alternatives that still meet consumers' expectations of indulgent, full bodied mouthfeel. Texture is not the only sensory characteristic for EPS to alter, and flavor and acidity are additional observable factors that may change. This ability of EPS to reduce syneresis and improve water retention helps keep the product stable, preventing the loss of volatile flavor compounds. EPS allows acids to be neutralized and reduces the amount of excessive acidification preventing unpleasant taste and increasing the palatability of fermented dairy products for consumers (Anumudu et al., 2024; Raj et al., 2022).

Probiotic Potential and Nutritional Enhancement

EPS can also improve the nutritional profile of yogurt and fermented milk. A few EPS produced by LAB have been reported to have potential prebiotic properties to stimulate beneficial gut microbiota. EPS improves the fermentation environment and increases the longevity of probiotic viability, both of which are correlated with the health benefits assumed from yogurt and other fermented dairy products. In addition, some EPS have been demonstrated to have bioactive effects, including antimicrobial and antioxidant properties, and thus may also contribute to endow consumers with additional health benefits (Abdalla et al., 2021; Andrew & Jayaraman, 2020). Incorporation of these functional properties enhances the nutritional value of yogurt and fermented milk products as effective protein and probiotic sources, and also as carriers of bioactive compounds that assist in promoting digestive health and general well-being.

Shelf Life and Stability

EPS also play a role in the stability and shelf life of yogurt and fermented milk. By their ability to improve consistency and prevent syneresis, the product will stay as it should for a long period, under storage. EPS presence minimizes moisture loss or separation changes of product due to which texture deterioration and flavors can occur. In particular, maintaining this stability is important to linear product quality for commercial yogurt production. EPS has been linked to enhanced stability in the presence of storage conditions with temperature fluctuations and changes in pH. Beyond maintaining the shelf life of yogurt during transportation and distribution, their role in ensuring the physical integrity of the product during distribution extends shelf life, reduces waste, and increases product availability to the consumer.

Innovation in Dairy Formulations

Areas of active innovation in the dairy industry relate to the use of EPS in yogurt and fermented milk products. Specific EPS-producing LAB can be used or fermentation conditions can be optimized to provide specific functional properties of EPS tailored to desired textures, improved stability, and overall sensory characteristics of the product. Furthermore, EPS biotechnology and fermentation technology are making significant advances, enabling the development of new dairy formulations such as low-fat, probiotic-rich, and functional dairy foods.

Cheese and Cheese Analogues

Texture and Firmness in Cheese

EPS plays a vital role in pronounced texture development in traditional cheese. The addition of EPS affects the gelation process of cheese making where milk proteins, mainly caseins become a network. They interact with casein micelles strengthening the protein network and improving the overall firmness and elasticity of the cheese. For example, early hardening is contributed to EPS produced by *L. lactis*, *S. thermophilus*, and Lactobacillus, especially in soft and semisoft cheese like mozzarella and ricotta. The hydrophilic networks formed by EPS trap water molecules and increase the water holding capacity (WHC) of cheese. It helps keep moisture in within the cheese matrix, a necessity to have that texture and not overly dry out. Such property is important in particular to cheese types such as mozzarella in which the ability to retain moisture is critical for having the perfect meltability and stretch ability.

Syneresis and Moisture Retention Control

EPS is important for reducing syneresis by increasing the water retention capacity of cheese and preventing whey separation during storage. EPS stabilize the gel structure as caseins form bond by binding water within their polysaccharide chains and so retard liquid loss and preserve the texture of the cheese. Particularly important is the ability of EPS to maintain moisture retention in cheese varieties in which there is prolonged aging otherwise the cheese could become excessively hard and brittle. EPS further prevents whey release from cheeses like feta, ricotta, and some soft cheeses to maintain the product's consistency and quality during the aging and storage process. EPS extends the shelf life of the cheese, improves the stability of the cheese under different temperature and humidity conditions, and helps the cheese maintain its desired texture and flavor profile when in distribution and consumption (Bihola et al., 2024).

Meltability and stretchability

In many instances, in processed cheeses and mozzarella in particular, important factors are meltability and stretchability. EPS modifies physical properties of the cheese such as viscosity and gel strength which then affect the meltability of the cheese (Bihola et al., 2024). Certain LAB strains produce polysaccharides of dextran and levan, that modify the molecular structure of the cheese matrix, facilitating better heat-induced melting (Talbi et al., 2023). However, the polysaccharides help stabilize the protein-lipid network such that when the cheese is heated there is a smooth, consistent melt. This is important because, for products such as pizza cheese, cheese needs to be able to melt and stretch without breaking up. In addition, EPS can enhance the stretchability of cheese through the creation of a more pliable, yet elastic, gel network, critical for cheeses such as mozzarella that are marketed for cooking applications that rely on elasticity. The presence of EPS increases the cheese's ability to stretch and develop a desirable stringy texture when melted, as well as greatly contributing to the desirable stringy texture of many pizza and pasta dishes.

Flavor Modification and Synergistic Effects

EPS is involved in the development of flavor in cheese production as it affects the microbial environment in fermentation. A few EPS have been shown to interact with some flavor precursors and thus contribute to the modification of the taste and aroma profiles of cheese. The release and retention of volatile compounds responsible for the characteristic flavors of cheese can be controlled through the structure and composition of EPS (K. Zhang et al., 2023). For example, EPS produced by LAB strains modulates fermentation hence production of aromatic compounds that contribute to the development of a more complex flavor profile is promoted. Similar to cheese analogues EPS can serve as an enhancer of flavor profile by providing a more natural sensory experience. Many of the cheese analogues are dairy-free, which can make it difficult to achieve the flavor characteristic of traditional cheese. EPS contributes to making dairy-free products with improved texture and mouthfeel, and may also contribute to keeping moisture and flavors in, stimulating of flavors of dairy-based cheeses.

EPS in Cheese Analogues

In plant-based cheese analogues, using primarily soy, almond, or coconut as the base, EPS is critical for the desired texture and consistency. Since cheese analogues don't have the natural casein and fatty oil content that dairy cheese has, they often have to have gelling agents and stabilizers added in order to get a similar mouth feel. As effective gelling agents, EPS can be used to firm up a texture that turns into a sliceable one, providing a mouthfeel reminiscent of traditional cheese. In addition, LAB or other microorganisms used in the fermentation of cheese analogues can also produce EPS that repairs water holding capacity and inhibits syneresis, as in dairy cheeses. Polysaccharides in these products help to ensure that cheese analogues stay moist during storage and use, prevent degradation and produce a more appealing texture when the product is actually used (Yang et al., 2020). EPS can also improve the meltability and stretchability of the product in cheese analogues.

When combined with EPS, the functionality of dairy cheese can be mimicked in plant-based cheese formulations, especially for texturing applications such as pizza or grilled cheese sandwiches.

Potential Health Benefits

Bioactive properties, such as antimicrobial, antioxidant, and immunomodulatory activities, have been observed for some of the EPS shared by both cheese fermentation and other processes. These properties can improve the nutritional value of traditional cheeses and cheese analogues similarly, with further health benefits to consumers. Specifically, some EPS are known to have prebiotic effects, that is, to stimulate the growth of beneficial gut microflora and to promote digestive health. And hence, incorporating EPS into cheese analogues leads to the functional benefits of these products as not only an alternative to dairy cheese but also functional foods that aid overall health (Korcz & Varga, 2021).

Cheese Formulation Innovation

The current increased demand for plant-based and functional foods has made the role of EPS in cheese analogues a more imperative area of interest. Major cheese analogue manufacturers use specific EPS producing microorganisms or add synthetic or natural EPS into cheese analogues, enabling the production of products with better texture, meltability, flavor, and overall sensory attributes. Moreover, improvements in EPS biotechnology and fermentation technology have led to the development of cheese analogues with even improved health benefits (i.e., lower fat content, higher fiber content), retaining the sensory characteristics that have made cheese so popular.

Ice Cream and Frozen Desserts

Contribute to Texture and Creaminess

EPS plays one of the most important roles in the texture and mouthfeel of ice cream and frozen desserts. Ice cream is thus a complex mixture of milk, cream, and sugar with air, which needs to be stabilized to get the desired creamy and smooth texture. Microorganisms, like Lactobacillus and Bifidobacterium, can produce EPS that acts as stabilizers by increasing the water-holding capacity of the mix and preventing ice crystal growth (chilling or freezing). Incorporation of EPS into ice cream formulations can impart a rich, creamy texture that closely mimics full-fat ice cream texture in reduced-fat or nondairy alternatives. The increase in the viscosity of the ice cream base may increase the creaminess of the product and its velvety texture, offering EPS the opportunity to enhance the product. In frozen desserts that don't often contain fat to contribute its a creamy mouthfeel, this is particularly important. Like dextrans and xanthan gum, EPS gives the body a greater presence than would otherwise occur, preventing a slimy or icy texture, and improving sensory experience.

Ice Crystal Formation Control

One of the problems with frozen desserts is the formation of large ice crystals, often in products that are low-fat, reduced-fat, or non-dairy (Romulo & Meindrawan, 2021). Too many ice crystals will give a gritty or icy texture to the consumer's sensory experience (Das & Hooda, 2023). Control of the ice crystal size during freezing is particularly reliant upon the convergence of the role played by EPS. The reason they provide a stabilizing function, to help reduce the rate of ice crystal formation and to keep the ice cream mixture from becoming bumpy, is because the ice crystal, that forms clogs up small spaces and is not so smooth or round that it can prevent them from coagualting together as well. This high water-binding capacity has been found to prevent large ice crystal formation by keeping water molecules inside polysaccharide chains. From freezing the ice cream then becomes smooth and homogeneous in texture. Its importance is particularly great in frozen desserts such as sorbets and nondairy ice creams, which present high water content and frequent ice crystal growth of large size occurs in them.

Viability of Viscosity and Mouthfeel Enhancement

Ice cream and frozen dessert product stability is, determined by their viscosity. By using EPS more viscosity of the product is achieved. Large branched molecular structures produce networks that trap water and stabilize the ice cream mixture, accounting for its viscosity-enhancing ability. The final product is a thicker consistency, which keeps both the product's mouthfeel and storage stability improved. Besides keeping the mouth structure in balance, EPS can enhance the sensory feeling of creaminess and smoothness when added to ice cream. It is most useful in frozen desserts that are low-fat or fat-free when the lack of fat might result in a less lavish texture. EPS can accurately mimic some of the textural properties of fat, which results in a richer more luxurious mouthfeel than a lightened version of fullfat ice cream while managing to stay calorie-free.

Water Holding Capacity and Syneresis Control

Retention of water and control of syneresis are critical control factors in producing high-quality frozen desserts. In other words, the release of water from the frozen matrix during storage is called syneresis and results in the formation of an undesirable liquid phase during storage. EPS also plays an important role in water retention by boosting the water binding in the ice cream, minimizing syneresis, and maintaining a stable, uniform texture throughout the shelf-life (Tan, 2019). EPS forms hydrophilic networks to prevent water migration from the ice cream structure and to prevent excess liquid formation on the surface of the frozen dessert. And that matters particularly in dairy-free frozen desserts where the water content is generally higher. In particular, guar gum and xanthan gum have shown to be particularly effective in controlling syneresis and preventing texture degradation during storage, as EPS.

Stability During Storage

When stored for a long period of time, ice cream and frozen desserts all change in texture, flavor and appearance. We cannot overemphasize the role of EPS in enhancing the stability of these compounds during freezing and storage. Therefore, these additives enable the improvement of viscosity, water retention, and regulation of ice crystal formation, and create a product that retains the desired texture and consistency over time. Together with the prevention of phase separation and the destabilization of emulsions in frozen dessert formulations, EPS also contributes to preventing phase separation and the destabilization of emulsions in frozen dessert formulations. EPS helps lend stability to the fat and water phases so that even after long-term storage, the texture continues to be smooth and creamy with little chance of the product hardening or crystallizing.

Contributions to NonDairy and ReducedFat Frozen Desserts

As the plant based and reduced fat frozen dessert market continues to grow, EPS have become indispensable in creating high-quality alternatives to traditional ice cream. Dairy-free frozen desserts made from plant-based ingredients like almond milk, soy milk, or coconut milk often don't have the fat and proteins found in dairy-based ice creams. These products also lack proteins and fat, but they rely on EPS to help with texture, mouthfeel, and stability, compensating for the gaps created. EPS is used in plant-based frozen desserts to replicate the smooth, creamy ice cream-like texture of dairy-based ice cream, enhancing the product sensory experience, and making these products more attractive to consumers (Syamala et al., 2024). For texture and sensory properties of full-fat ice cream to be maintained, EPS allows the creaminess and smoothness to be retained without increasing calorie content.

Sensory Attributes and Flavor Retention

While EPS serves functional roles, it also affects sensory attributes and flavor retention of ice cream and frozen desserts. EPS improves the texture and eating experience by improving water retention, giving our dishes a greater mouthfeel. Additionally, they are able to stabilize emulsions and decrease syneresis, thus preventing volatile flavor compounds from leaching out of the product as it is sold and stored, which maintains its flavor profile throughout the useful life of the product. EPS can provide the retained flavor profile in nondairy frozen desserts in which the lack of fat or dairy protein proteins can lead to flavor distortion. By offering a stable matrix, the EPS give better flavor release and a better sensory experience, making these products more attractive to consumers looking for alternatives to conventional meat.

Frozen Dessert Formulations

With the increasing demand for healthier and innovative frozen desserts, EPS is also emerging to be used in the formulation of new products. The development of various frozen desserts of acceptable texture, sensory attributes, and nutritional profiles is enabled by improvements in EPS technology and fermentation processes. EPS enable manufacturers to formulate dairy based or plant-based products with improved functionality, attractive textures, and even quality.

Dairy Based Sauces and Dressings

Texture, Creaminess

EPS can add texture and mouthfeel to dairy-based sauces and dressings, one of the main roles they play in the industry. Cream-based sauces, salad dressings, and dips to feature a smooth, rich texture; many dairy sauces and dressings do too. This can be achieved with polysaccharides, which enhances viscosity and develops a gel-like structure in the product, giving the product a creamy and cohesive character (Asase & Glukhareva, 2024). Specifically, in low-fat or fat-free formulations, in the lack of natural creaminess, this is of greatest importance. Among the EPS (xanthan gum, guar gum) that function as fat substitutes and thickeners such as in sauce or dressings, they are emulsifying and make the sauce or dressing luxurious but not fattening (Pettitt, 2020). EPS also can improve the smoothness of mixtures in dairy-based dressings to prevent the formation of lumps or separation of ingredients. This provides a stable and homogeneous product with easy applicability that is also visually appealing. The polysaccharides' engagement provides the capability to improve the texture and thus improve the ability of sauces and dressings to coat the food evenly giving an improved sensory experience.

Viscosity Regulation

The critical properties in the formulation of sauces and dressings is viscosity, other properties include pourability and spread ability. Interacting with water and other ingredients in the formulation, EPS can then control the viscosity of a dairy-based sauce. When EPS is added to a dairy based sauce the mixture becomes thicker and has a desirable thickness, neither too runny nor too thick. Particularly important in salad dressings, a high degree of control over the viscosity of this mixture is desired to achieve a pourable thickness that ensures an even distribution over salad leaves. EPS contributes to stable sauce consistency in such cream-based sauces as Alfredo or béchamel to be maintained over their shelf life. EPS improves the water binding capacity, prevents excessive thinning, and helps keep the sauce together, even after many heating or refrigeration cycles.

Emulsification and Stability

In the realm of dairy-based sauces and dressings, the emulsification of fats and water is usually referred to as an unstable process. When there is no proper stabilization, the fat phase can separate from the water phase and an undesired product texture and appearance can be obtained. EPS serves as an emulsifier and increases the stability of these emulsions to help prevent phase separation during storage. Gellan, xanthan gum, dextrans, and other EPS have good emulsifying properties that stabilize fat and water phases in dairy sauces and dressings (Suryawanshi et al., 2022). These polysaccharides may serve to keep the fat droplets apart by interacting with the water phase and forming a network gel (stabilizing the emulsion and preventing the coalescence of the fat droplets). Like many polymer powders, EPS is particularly useful to prevent the oil or cream phase of dressings from separating, producing a smoother, more consistent product.

Water Holding Capacity and Absence of Syneresis

The quality and stability of dairy-based sauces and dressings are critically dependent on water-holding capacity. The water retention ability of the product is improved with EPS to decrease the risk of syneresis (expulsion of liquid) upon storage. It's especially important in creamy sauces and dressings where the release of water can give an undesirable watery phase on the surface. EPS plays into the network of the sauce or dressing to trap water molecules and prevent them from getting separated. It reduces syneresis and prohibits the existence of syneresis, permitting storage time with product stability by maintaining its quality and preventing unwanted liquid formation. But, importantly, EPS has highly desirable water retention properties which help create a dressing or sauce that will maintain its desired mouthfeel and consistency after refrigeration or long-term storage (Werning et al., 2022).

Reduced Calorie and Low-Fat formulations

In response to the demand for healthier alternatives to traditional high-fat dairy products, low-fat and reduced-calorie dairy-based sauces and dressings have been developed. The reduced fat content adversely affects the texture, viscosity, and mouthfeel of the products, depending on the level of fat, and EPS plays a key role in product formulation by improving texture, viscosity, and mouthfeel. With its creamy, rich appealing sensory attributes and its relative healthiness, EPS may help to replicate the creamy, rich consistency provided by fat, without compromising health (Yildiz & Karatas, 2018). EPS, such as guar gum, xanthan gum, and carrageenan, are used in fat-reduced or fat-free formulations to provide mouthfeel or thickness where fat would usually have been used. However, they ensure a lower caloric content and help create the creaminess and smooth texture consumers expect from dairy sauces and dressings. Since the functionality enables manufacturers to make reduced-fat versions of products such as creamy ranch dressing or béchamel sauce with no loss of texture or sensory appeal this functionality is beneficial to manufacturers.

Sensory and Flavor Retention

Beyond their texture and viscosity functions, EPS appears to have a positive impact on flavor retention and the mouthfeel of dairy-based sauces and dressings. EPS can form a gel-like network around the product that allows some flavor compounds that might otherwise dissipate or degrade during processing or storage to be better retained. Additionally, the thickening properties of EPS make the flavor more evenly distributed throughout the sauce or dressing. As a result, the taste remains constant and the flavor sticks to the food.

Dairy-Based Sauces and Dressings Innovation

Dairy-based sauces and dressings continue to shift consumer preferences towards healthier and more innovative choices, making the role of EPS in dairy-based sauces and dressings even more important. The improved texture, improved stability and lower calorie content of sauces and dressings are possible through further development of EPS technology and new types of polysaccharides. The relevance of these innovations is particularly significant in the manufacture of plant-based or dairy-free alternatives to conventional dairy sauces where EPS can simulate dairy product flavor without using animal-based ingredients.

HEALTH BENEFITS OF EPS IN DAIRY PRODUCTS

Probiotic Properties and Gut Health

Modulation of Gut Microbiota

Modulation of the gut microbiota, however, is one of the most important probiotic properties of EPS. The human gastrointestinal tract has trillions of microorganisms, many of which are beneficial bacteria that participate in food digestion, synthesis of essential nutrients, and supporting intestinal homeostasis. EPS produced by LAB can act as prebiotics, substrates that selectively enhance the growth or activity of beneficial microflora residing in the gut. EPS can help maintain intestinal health by selectively boosting the growth of beneficial bacteria such as Bifidobacteria and Lactobacilli and dysbiosis pathogen development (Oerlemans et al., 2021). The composition of EPS, which can include diverse sugar residues of glucose, galactose, and rhamnose, leads to its largely prebiotic effect attributed to their composition. This selected fermentation allows for beneficial microorganisms to grow and also create short-chain fatty acids (SCFAs), which provide a range of health benefits, including anti-inflammatory and repair of GI tract.

Integrity and Protection of Gut tract

Maintaining the body's integrity by protecting it from harmful pathogens and maintaining good health requires the integrity of the intestinal barrier. However, intestinal epithelium serves to be a protective barrier against the in-translocation of pathogens and toxins from the gut to the bloodstream. Stimulation of mucus production by EPS has been demonstrated to improve the gut barrier function by enhancing the protection of the intestinal lining (Yang et al., 2023). LAB strains (genera Lactobacillus and Bifidobacterium) have been reported to promote EPS mucus secretion, generating a thicker and more turgid mucus layer that can protect the epithelium from harmful substances. Aside from the ability to increase mucus production, EPS may also help stabilize the tight layer of cells that make up the intestine. Proteins in the cells that make up the boundary of the intestinal epithelium are called tight junctions, they help keep harmful molecules from leaking out of the boundary. EPS can lead to the strengthening of these tight junctions, and thereby prevent intestinal permeability (also known as "leaky gut," the passage of harmful bacteria and toxins into the bloodstream). On the basis of gut barrier dysfunction in conditions characterized with impaired gut barrier function, such as irritable bowel syndrome (IBS) and inflammatory bowel diseases (IBD), this property is of particular interest. It can also neutralize heavy metal toxicity affects (Dahiya et al., 2024)

Anti-Inflammatory Effects

One mechanism by which EPS has been shown to have anti-inflammatory effects is gut health, and inflammation can be reduced in the GI tract by reducing inflammation. EPS has been shown to modulate the immune response through the regulation of the production of proinflammatory cytokines and the production of anti-inflammatory molecules. For instance, Lactobacillus and Bifidobacterium strain-derived EPS, are known to reduce expression of proinflammatory cytokines such as TNFα, IL6, and IL1β in the gut (Gu et al., 2023). These EPS, however, stimulate the secretion of anti-inflammatory cytokines (such as IL10), which can regulate inflammation and maintain a healthy immune balance (Yue et al., 2023). It can also help reduce inflammatory bowel disease, and the symptoms of other inflammatory intestinal disorders as well as improve gut comfort and prevent chronic gut disease.

Helping improve Digestive Health and Motility

EPS also has an effect on GI motility and on digestion. Specific EPS has been demonstrated to stimulate peristalsis (wavelike muscle contractions that propel food through the digestive tract) and thus improve bowel movements as well as decrease the chance of constipation (T. Zhang et al., 2023). EPS may help reduce discomfort from sluggish digestion and promote regularity by boosting gut motility. Besides, LAB resultant EPS will be stressed and pave the way for digestive enzyme composition that can help to break down nutrients and facilitate nutrient absorption. EPS can help with overall digestive health through increased efficiency of nutrient digestion and absorption.

Mental Health and Gut Brain Axis

There is growing evidence that modulation of the gut microbiota by EPS contributes to the modulation of brain function and regulation of mood. The capacity of EPS to modulate gut microbiota composition, generate bioactive metabolites such as SCFAs, and attenuate systemic inflammation may have implications for mental health problems, such as anxiety, depression, and stress. LAB strains are capable of producing EPS that can exert influence over the gut-brain axis via modulation of neurotransmitter secretion that is dominated by gut-derived serotonin. EPS may indirectly positively affect mental well-being by improving gut health, a first indication that this kind of movement-associated ingredients can be functional ingredients not only for the gut but also for mental health support.

Immune Modulation

Interaction with Immune Cells

Interaction of EPS with immune cells such as macrophages, dendritic cells, and T lymphocytes is one of the important mechanisms by which EPS modulates the immune system (Chaisuwan et al., 2020; Jaseera & Abdulla, 2021). Detected pathogens and triggers of immune responses rely on these immune cells. By binding itself to PRRs, e.g., TLRs, EPS can activate immune cells. Pathogen-associated molecular patterns (PAMPs) and other microbial elicitors trigger immune responses triggered by PRR that recognize PAMPs and other microbial signals. Stimulation of macrophages and dendritic cells by EPS utilizes this interaction to promote the release of important proinflammatory cytokines, including TNFα, IL6, and IL1β, which are important in the initiation of the innate immune response. All of this helps them to also help hold the immune system's response in check so it doesn't respond in a way that leads to too much inflammation. This balance of immune activation may also help keep the immune system in check and lower your risk for autoimmune diseases.

Stimulation of Gut Associated Lymphoid Tissue (GALT)

The GALT is a large part of the immune system as it lies at the induction site of immune responses to gastrointestinal pathogens. Also, EPS contains proteins that promote the body's mucosal immune defenses, predominantly in GALT upon ingestion of probiotic strains belonging to Lactobacillus and Bifidobacterium. The finding that there are different signaling pathways preferred by EPS for its interaction with the epithelial cells of the gut makes sense, as they have to be able to activate immune responses in this region, such as secretion of secretory IgA, an antibody that defends the mucosal surface from microbial invasion (Chen et al., 2020). Stimulation of sIgA production by EPS is important since sIgA is the first line of defense against pathogens in the gastrointestinal tract (Li et al., 2020). In addition, the gut microbiota composition and function, in turn regulating GALT activation and systemic immune responses are influenced by EPS. EPS is a source of these mechanisms that support overall immune function, maintain gut health, and guard against enteric infections.

Cytokine Production Modulation

Cytokine production in the local and systemic immune responses is influenced by EPS. Immune modulation is largely mediated by the production of cytokines as these signaling molecules modulate the activity of several immune cells including T-helper cells, B cells, and natural killer (NK) cells. Stimulating cytokine production by EPS can help to boost the immune response to the infection, but if not excessive, to prevent overreaction of the immune system creating excessive inflammation. Some even show the ability to stimulate the production of proinflammatory cytokines such as TNFα and IL1β to activate immune cells and promote clearance of pathogens. However, like the ER, inflammation-stimulating signals that are released by the EPS can also trigger the release of anti-inflammatory cytokines, including IL10, that act to regulate the immune response and dampen the destructive process of chronic inflammation. This balance between pro-inflammatory and anti-inflammatory cytokines must remain in check to maintain an optimum balance of the immune system, and to prevent immune-mediated diseases like allergies and autoimmune disorders.

Adaptive Immunity Modulation

among these include Th1 cells that are needed to fight intracellulars pathogens and Th2 cells involved with responses to extracellular parasites and allergens (Lekki-Jóźwiak & Bąska, 2023). Tailored immune responses resulting from EPS-mediated effects can tip than Th cell balance, favoring specific responses to different types of pathogens to better fight infections and maintain immune tolerance. In addition, EPS was shown to bind Treg, and enhances the proliferation of Tregs, a subset of immune cells that are critical for maintaining immune tolerance and protection from autoimmunity. Treg number increase could come from EPS and could help prevent immune system illness and decrease the inflammation levels of autoimmune disease and chronic conditions such as rheumatoid arthritis and multiple sclerosis.

Modulation of Inflammation

Chronic low grade inflammation is associated with many diseases, ranging from cardiovascular diseases to metabolic syndrome and IBD. Some of the effects might mitigate damage attributed to chronic inflammation on health. The suggestion that EPS can influence immune cell activity and cytokine production, reducing proinflammatory mediators and augmenting anti-inflammatory mediators to end inflammation is made. Lactobacillus and Bifidobacterium species-derived EPS have been demonstrated to decrease proinflammatory cytokines and prevent NFκB pathway activation (Nowak et al., 2019). The inhibition of these inflammatory signaling pathways contributes to prevent inflammation escalation and reduce the risk of inflammatory disease. In addition, EPS may promote the production of molecules that inhibit the inflammatory response, including IL10, and can help tissue repair.

Immune Related Disease Potential

Its immune modulatory properties have led EPS to be considered a possible therapeutic agent for immune related diseases. They are very promising in allergic, asthma, and IBD, and rheumatoid arthritis.EPS are being sought to develop functional foods/ supplements which modulate immune function or prevent or manage immune mediated diseases due to its ability to regulate immune responses and decrease chronic inflammation.ity to regulate immune responses and to decrease chronic inflammation, EPS are ideal for the development of functional foods or supplements that modulate immune function or prevent or manage immune-mediated diseases. In autoimmune diseases such as rheumatoid arthritis and lupus, whose immune systems turn against their healthy tissues, immunity itself might be controlled by EPS to change the behavior of immune cells and to induce immune tolerance. Rearch might shorten the suffering of these diseases by reducing over inflammation and promoting the function of the regulatory T cell.

EPS in Cancer Immunotherapy

In addition, it had recently been implicated in cancer immunotherapy: as a powerful molecule that can help the body's immune surveillance and expand its ability to identify and target tumor cells. EPS derived from LAB strains has been shown to stimulate immune cells like macrophage and dendritic cells. EPS may activate immune cells in the body to help boost the body's detection and elimination of cancer cells. Additionally, EPS can modulate the tumor microenvironment via inflammatory response, activation of antitumor immunity, and abatement of tumor-induced immunosuppression (Jiang et al., 2024). Specialized in curbing immune cells that become associated with cancer activity, EPS could help make these treatments more effective, while also helping to bolster the body's innate effort to battle the disease.

Role in Reducing Lactose Intolerance

EPS also plays an important role to promote the growth of beneficial gut bacteria which aid in lactose digestion. Often referred to as probiotics, these bacteria are often able to make lactase-like enzymes that partially hydrolyze lactose (Jurášková et al., 2022). This microbial fermentation of lactose can lessen the strain on the host's digestive system reducing symptoms of lactose malabsorption and increasing lactose tolerance. Additionally, EPS promote the entire health of the gut barrier function. They cause the intestines to manufacture mucus to protect the epithelial lining and to prevent unabsorbed lactose from irritating. It can even provide a protective effect against bloat and discomfort which is common in people who are lactose intolerant. In fact, EPS also contribute to the quality of the gut barrier, and integrity of the gut barrier is important for the appropriate digestion and absorption of nutrients, especially lactose. In addition, EPS also have anti-inflammatory properties and thus more than mitigate the symptoms of lactose intolerance. In those with lactose malabsorption, undigested lactose can cause gut inflammation, making matters even worse. It can modulate immune responses, reduce inflammatory cytokines, providing help to soothe the digestive tract, easing discomfort and generally improving gut health.

TECHNOLOGICAL CHALLENGES AND CONSIDERATIONS

Optimization of EPS Production in Dairy Processing

Extracellular polysaccharides (EPS) production in dairy processing imposes challenges for the highest yield with the required proper functional product properties. The production of EPS is a function of fermentation conditions (temperature, pH and oxygen availability and culture medium composition) and due to the strain of lactic acid bacteria used (LAB) (Nguyen et al., 2020). EPS from LAB is differentiated by a range of characteristics (viscosity, molecular weight, solubility) which then determine the sensory and texture properties of dairy products. Therefore, the optimal balance between yield of EPS and quality achieved is achieved by selection of bacterial strains suitable for a given application for example use in preparing yogurt, cheese or any other fermented dairy product, and fermentation conditions.

In addition, ecological and fermentation control difficulties may not let high productions of EPS in dairy processing; proper bioreactor design to produce EPS quality and quantity is needed. Therefore, the growth medium to be used in dairy fermentation should be optimized to allow high EPS production with no loss of taste and nutritional value of the product produced. Often, this optimization involves finetuning fermentation parameters and including some precursors or cofactors in the medium to improve EPS yield. In addition, post-fermentation processes, including separation, purification, and incorporation into dairy products, also present problems since EPS stability and functionality have to be maintained during processing and storage.

Stability and Shelf-Life Implications

The stability and shelf life of extracellular polysaccharides (EPS) are important for their functionality in dairy products and other food applications. Lactic acid bacteria (LAB) produce EPS which is used in different dairy products for their textural, sensory, and functional benefits. Nevertheless, factors such as temperature, pH, or other ingredients can influence their stability. For refrigerated storage, EPS is generally stable, although prolonged exposure to high temperatures or variable conditions can cause deterioration of or loss of functional properties of EPS. Heat-sensitive EPS may also hydrolyze or suffer a structural breakdown, impairing their ability to improve the texture, viscosity, or water retention in processed food products. For good EPS storage conditions especially for heat-treated products like yogurt and cheese, it is necessary. In addition, the shelf life of EPS can be dependent upon their interaction with other ingredients contained in food products. For example, high sugar or salt concentrations will impart EPS stability while acid conditions can degrade EPSs (Babiak & Krzemińska, 2021; Z.-y. Yin et al., 2024). Therefore, manufacturers must formulate the EPS, and store them appropriately such that they keep the most desirable properties over the shelf life of the product.

Regulatory Aspects and Labeling Requirements

All these efforts by. organizations aim to regulate the use of extracellular. polysaccharides (EPS) in food products, mainly dairy, guidelines for the approval and use of EPS as food additives or ingredients have been set by regulatory bodies such as the Food and Drug Administration (FDA) in the United States and the European Food Safety Authority (EFSA) in Europe. By these guidelines, EPS is allowed for use in food products only if tested for safety (i.e., toxicological studies, allergenicity potential, and microbiological quality). EPS that derived from microorganisms known to be generally safe such as Lactobacillus and Bifidobacterium are generally considered safe, but all products must also be adhered to the authorities' laws regarding product. According to food labeling regulations, EPS in food products must be disclosed clearly, and any related claim regarding EPS must also be listed.

Food products in the EU and the US where EPS is added are most commonly treated according to the EU Food Additives or functional ingredients, the latter requires accurate labeling. The specific EPS used must appear on the product label (and if necessary, its European number or scientific name). Moreover, if the EPS possesses the potential to be allergenic or possesses known specific health benefits (regulatory health claims), such claims must be supported and comply with further health claim regulations. This makes product ingredient information transparent and enables the consumer to know what they are eating.

CONCLUSION

EPS are important dairy processing components that provide numerous functional benefits to enhance dairy product quality and texture. Due to their ability to improve viscosity, water retention, mouthfeel, and syneresis control, they have an important role in the production of products such as yogurt, cheese, ice cream, and fermented milk. In addition, EPS obtained from lactic acid bacteria (LAB) promotes additional health benefits, including probiotic and immune modulation adding to the demand for functional foods. Main challenges faced during industrialization of EPS are strain choice, fermentation conditions, and the retention of functional properties. New biotechnological and fermentation technologies facilitate production of EPS. EPS will remain subject of research in future and promises beneficial processing and health benefits.

REFERENCES

Abdalla, A. K., Ayyash, M. M., Olaimat, A. N., Osaili, T. M., Al-Nabulsi, A. A., Shah, N. P., & Holley, R. (2021). Exopolysaccharides as antimicrobial agents: Mechanism and spectrum of activity. *Frontiers in Microbiology*, *12*, 664395. DOI: 10.3389/fmicb.2021.664395 PMID: 34093478

Akan, G., & Oner, E. T. (2022). Gel properties of microbial polysaccharides. In *Polysaccharides of Microbial Origin: Biomedical Applications* (pp. 579-598). Springer. DOI: 10.1007/978-3-030-42215-8_34

Andrew, M., & Jayaraman, G. (2020). Structural features of microbial exopolysaccharides in relation to their antioxidant activity. *Carbohydrate Research*, *487*, 107881. DOI: 10.1016/j.carres.2019.107881 PMID: 31805426

Angelin, J., & Kavitha, M. (2020). Exopolysaccharides from probiotic bacteria and their health potential. *International Journal of Biological Macromolecules*, *162*, 853–865. DOI: 10.1016/j.ijbiomac.2020.06.190 PMID: 32585269

Anumudu, C. K., Miri, T., & Onyeaka, H. (2024). Multifunctional Applications of Lactic Acid Bacteria: Enhancing Safety, Quality, and Nutritional Value in Foods and Fermented Beverages. *Foods*, *13*(23), 3714. DOI: 10.3390/foods13233714 PMID: 39682785

Appiani, M., Cattaneo, C., & Laureati, M. (2023). Sensory properties and consumer acceptance of plant-based meat, dairy, fish and eggs analogs: A systematic review. *Frontiers in Sustainable Food Systems*, *7*, 1268068. DOI: 10.3389/fsufs.2023.1268068

Arab, M., Yousefi, M., Khanniri, E., Azari, M., Ghasemzadeh-Mohammadi, V., & Mollakhalili-Meybodi, N. (2023). A comprehensive review on yogurt syneresis: Effect of processing conditions and added additives. *Journal of Food Science and Technology*, *60*(6), 1656–1665. DOI: 10.1007/s13197-022-05403-6 PMID: 37187980

Asase, R. V., & Glukhareva, T. V. (2024). Production and application of xanthan gum—prospects in the dairy and plant-based milk food industry: A review. *Food Science and Biotechnology*, *33*(4), 749–767. DOI: 10.1007/s10068-023-01442-7 PMID: 38371690

Atik, D. S., Öztürk, H. İ., & Akın, N. (2024). Perspectives on the yogurt rheology. *International Journal of Biological Macromolecules*, *263*, 130428. DOI: 10.1016/j.ijbiomac.2024.130428 PMID: 38403217

Babiak, W., & Krzemińska, I. (2021). Extracellular polymeric substances (EPS) as microalgal bioproducts: A review of factors affecting EPS synthesis and application in flocculation processes. *Energies*, *14*(13), 4007. DOI: 10.3390/en14134007

Basiri, S. (2021). Applications of microbial exopolysaccharides in the food Industry. *Avicenna Journal of Medical Biochemistry*, *9*(2), 107–120. DOI: 10.34172/ajmb.2021.16

Berthold-Pluta, A. M., Pluta, A. S., Garbowska, M., & Stasiak-Różańska, L. (2019). Exopolysaccharide-producing lactic acid bacteria–health-promoting properties and application in the dairy industry. *Postepy Mikrobiologii*, *58*(2), 191–204. DOI: 10.21307/PM-2019.58.2.191

Bihola, A., Sharma, H., Chaudhary, M., Bumbadiya, M., Kumar, D., & Adil, S. (2024). Recent developments in cheese technologies. *Food Reviews International*, 1–35.

Chaisuwan, W., Jantanasakulwong, K., Wangtueai, S., Phimolsiripol, Y., Chaiyaso, T., Techapun, C., Phongthai, S., You, S. G., Regenstein, J. M., & Seesuriyachan, P. (2020). Microbial exopolysaccharides for immune enhancement: Fermentation, modifications and bioactivities. *Food Bioscience*, *35*, 100564. DOI: 10.1016/j.fbio.2020.100564

Chen, K., Magri, G., Grasset, E. K., & Cerutti, A. (2020). Rethinking mucosal antibody responses: IgM, IgG and IgD join IgA. *Nature Reviews. Immunology*, *20*(7), 427–441. DOI: 10.1038/s41577-019-0261-1 PMID: 32015473

Dahiya, P., Kumari, S., Behl, M., Kashyap, A., Kumari, D., Thakur, K., & Walia, A. (2024). Guardians of the Gut: Harnessing the Power of Probiotic Microbiota and Their Exopolysaccharides to Mitigate Heavy Metal Toxicity in Human for Better Health. *Probiotics and Antimicrobial Proteins*, *16*(6), 1–17. DOI: 10.1007/s12602-024-10281-9 PMID: 38733461

Das, N., & Hooda, A. (2023). Chemistry and Different Aspects of Ice Cream. In *The Chemistry of Milk and Milk Products* (pp. 65–86). Apple Academic Press. DOI: 10.1201/9781003340706-5

Duboux, S., Van Wijchen, M., & Kleerebezem, M. (2021). The possible link between manufacturing and probiotic efficacy; a molecular point of view on Bifidobacterium. *Frontiers in Microbiology*, *12*, 812536. DOI: 10.3389/fmicb.2021.812536 PMID: 35003044

Goh, K. K., Teo, A., Sarkar, A., & Singh, H. (2020). Milk protein-polysaccharide interactions. In *Milk proteins* (pp. 499–535). Elsevier. DOI: 10.1016/B978-0-12-815251-5.00013-X

Gu, Z., Meng, S., Wang, Y., Lyu, B., Li, P., & Shang, N. (2023). A novel bioactive postbiotics: From microbiota-derived extracellular nanoparticles to health promoting. *Critical Reviews in Food Science and Nutrition*, *63*(24), 6885–6899. DOI: 10.1080/10408398.2022.2039897 PMID: 35179102

Han, H.-S., Shin, J.-S., Song, Y.-R., Rhee, Y. K., Cho, C.-W., Ryu, J. H., Inn, K.-S., Hong, H.-D., & Lee, K.-T. (2020). Immunostimulatory effects of polysaccharides isolated from young barley leaves (Hordeum vulgare L.) with dual activation of Th1 and Th2 in splenic T cells and cyclophosphamide-induced immunosuppressed mice. *International Journal of Biological Macromolecules*, *147*, 954–964. DOI: 10.1016/j.ijbiomac.2019.10.062 PMID: 31730956

Jaseera, K., & Abdulla, T. (2021). Microbial EPS as Immunomodulatory Agents. *Microbial Exopolysaccharides as Novel and Significant Biomaterials*, 235-264.

Jiang, B., Wei, M., He, Y., Wang, Y., Zhang, Q., Wei, H., & Tao, X. (2024). Antitumor effect of exopolysaccharide from Lactiplantibacillus plantarum WLPL09 on melanoma mice via regulating immunity and gut microbiota. *International Journal of Biological Macromolecules*, *254*, 127624. DOI: 10.1016/j.ijbiomac.2023.127624 PMID: 37918610

Jurášková, D., Ribeiro, S. C., & Silva, C. C. (2022). Exopolysaccharides produced by lactic acid bacteria: From biosynthesis to health-promoting properties. *Foods*, *11*(2), 156. DOI: 10.3390/foods11020156 PMID: 35053888

Korcz, E., & Varga, L. (2021). Exopolysaccharides from lactic acid bacteria: Techno-functional application in the food industry. *Trends in Food Science & Technology*, *110*, 375–384. DOI: 10.1016/j.tifs.2021.02.014

Kumar, S., Kumar, S., Bumbadia, M., & Singh, S. K. (2022). Prospects and functionality of bacterial exopolysaccharides in dairy foods: A Review: Bacterial exopolysaccharides in dairy. *Journal of AgriSearch*, *9*(1), 1–5. DOI: 10.21921/jas.v9i01.9884

Leal, M. R. S., Albuquerque, P. B. S., Rodrigues, N. E. R., dos Santos Silva, P. M., de Oliveira, W. F., dos Santos Correia, M. T., & Coelho, L. C. B. B. (2024). A review on the use of polysaccharides as thickeners in yogurts. *Carbohydrate Polymer Technologies and Applications*, *8*, 100547. DOI: 10.1016/j.carpta.2024.100547

Lekki-Jóźwiak, J., & Bąska, P. (2023). The Roles of Various Immune Cell Populations in Immune Response against Helminths. *International Journal of Molecular Sciences*, *25*(1), 420. DOI: 10.3390/ijms25010420 PMID: 38203591

Li, Y., Jin, L., & Chen, T. (2020). The effects of secretory IgA in the mucosal immune system. *BioMed Research International*, *2020*(1), 2032057. DOI: 10.1155/2020/2032057 PMID: 31998782

Lu, X., Chen, J., Guo, Z., Zheng, Y., Rea, M. C., Su, H., Zheng, X., Zheng, B., & Miao, S. (2019). Using polysaccharides for the enhancement of functionality of foods: A review. *Trends in Food Science & Technology*, *86*, 311–327. DOI: 10.1016/j.tifs.2019.02.024

Muninathan, C., Guruchandran, S., Viswanath Kalyan, A. J., & Ganesan, N. D. (2022). Microbial exopolysaccharides: Role in functional food engineering and gut-health management. *International Journal of Food Science & Technology*, *57*(1), 27–34. DOI: 10.1111/ijfs.15334

Nguyen, P.-T., Nguyen, T.-T., Bui, D.-C., Hong, P.-T., Hoang, Q.-K., & Nguyen, H.-T. (2020). Exopolysaccharide production by lactic acid bacteria: The manipulation of environmental stresses for industrial applications. *AIMS Microbiology*, *6*(4), 451–469. DOI: 10.3934/microbiol.2020027 PMID: 33364538

Nowak, A., Paliwoda, A., & Błasiak, J. (2019). Anti-proliferative, pro-apoptotic and anti-oxidative activity of Lactobacillus and Bifidobacterium strains: A review of mechanisms and therapeutic perspectives. *Critical Reviews in Food Science and Nutrition*, *59*(21), 3456–3467. DOI: 10.1080/10408398.2018.1494539 PMID: 30010390

Oerlemans, M. M., Akkerman, R., Ferrari, M., Walvoort, M. T., & de Vos, P. (2021). Benefits of bacteria-derived exopolysaccharides on gastrointestinal microbiota, immunity and health. *Journal of Functional Foods*, *76*, 104289. DOI: 10.1016/j.jff.2020.104289

Pettitt, D. J. (2020). Xanthan gum. In *Food hydrocolloids* (pp. 127–149). CRC Press.

Pourjafar, H., Ansari, F., Sadeghi, A., Samakkhah, S. A., & Jafari, S. M. (2023). Functional and health-promoting properties of probiotics' exopolysaccharides; isolation, characterization, and applications in the food industry. *Critical Reviews in Food Science and Nutrition*, *63*(26), 8194–8225. DOI: 10.1080/10408398.2022.2047883 PMID: 35266799

Prasad, S., & Purohit, S. R. (2023). Microbial exopolysaccharide: Sources, stress conditions, properties and application in food and environment: A comprehensive review. *International Journal of Biological Macromolecules*, *242*, 124925. DOI: 10.1016/j.ijbiomac.2023.124925 PMID: 37236568

Prete, R., Alam, M. K., Perpetuini, G., Perla, C., Pittia, P., & Corsetti, A. (2021). Lactic acid bacteria exopolysaccharides producers: A sustainable tool for functional foods. *Foods*, *10*(7), 1653. DOI: 10.3390/foods10071653 PMID: 34359523

Raj, T., Chandrasekhar, K., Kumar, A. N., & Kim, S.-H. (2022). Recent biotechnological trends in lactic acid bacterial fermentation for food processing industries. *Systems Microbiology and Biomanufacturing*, *2*(1), 14–40. DOI: 10.1007/s43393-021-00044-w

Romulo, A., & Meindrawan, B. (2021). Effect of dairy and non-dairy ingredients on the physical characteristic of ice cream. IOP Conference Series: Earth and Environmental Science. PMID: 34845955

Surber, G., Schäper, C., Wefers, D., Rohm, H., & Jaros, D. (2021). Exopolysaccharides from Lactococcus lactis affect manufacture, texture and sensory properties of concentrated acid milk gel suspensions (fresh cheese). *International Dairy Journal*, *112*, 104854. DOI: 10.1016/j.idairyj.2020.104854

Suryawanshi, N., Naik, S., Eswari, J. S., Nazir, F., Saeed, M., Abbas, A., . . . Nasir, M. (2022). Food science and applied biotechnology. *Foo d Science and Applied Biotechnology, 2022.*

Syamala, A., Dubey, K., & Salunke, P. (2024). Milk and Dairy Products Analogues. In *Food Analogues: Emerging Methods and Challenges* (pp. 35-74). Springer. DOI: 10.1007/978-3-031-69872-9_2

Talbi, C., Elmarrkechy, S., Youssfi, M., Bouzroud, S., Belfquih, M., Sifou, A., . . . Bouyahya, A. (2023). Bacterial exopolysaccharides: from production to functional features. *Progress In Microbes & Molecular Biology, 6*(1).

Tan, J. (2019). Structuring semisolid foods. *Rheology of Semisolid Foods*, 167-201.

Tiwari, S., Kavitake, D., Devi, P. B., & Shetty, P. H. (2021). Bacterial exopolysaccharides for improvement of technological, functional and rheological properties of yoghurt. *International Journal of Biological Macromolecules*, *183*, 1585–1595. DOI: 10.1016/j.ijbiomac.2021.05.140 PMID: 34044028

Tribby, D., & Teter, V. (2023). Yogurt. In *The sensory evaluation of dairy products* (pp. 199–234). Springer. DOI: 10.1007/978-3-031-30019-6_8

Wang, L., Gu, Y., & Lv, Z. (2022). Processing properties of yogurt as affected by the EPS produced by Leuconostoc mesenteroides XR1. *International Journal of Food Science & Technology*, *57*(7), 4076–4085. DOI: 10.1111/ijfs.15722

Wang, L., Tian, H., Liu, W., Zheng, H., Wu, H., Guan, Y., Lu, Q., & Lv, Z. (2023). Effects of EPS-producing Leuconostoc mesenteroides XR1 on texture, rheological properties, microstructure and volatile flavor of fermented milk. *Food Bioscience, 56*, 103371. DOI: 10.1016/j.fbio.2023.103371

Werning, M. L., Hernández-Alcántara, A. M., Ruiz, M. J., Soto, L. P., Dueñas, M. T., López, P., & Frizzo, L. S. (2022). Biological functions of exopolysaccharides from lactic acid bacteria and their potential benefits for humans and farmed animals. *Foods, 11*(9), 1284. DOI: 10.3390/foods11091284 PMID: 35564008

Xu, Q., Wang, H., Ren, Y., Sun, M., Zhang, T., Li, H., & Liu, X. (2024). Functionality and application of emulsion gels in fat replacement strategies for dairy products. *Trends in Food Science & Technology, 152*, 104673. DOI: 10.1016/j.tifs.2024.104673

Xu, Z., Guo, Q., Zhang, H., Xiong, Z., Zhang, X., & Ai, L. (2021). Structural characterisation of EPS of Streptococcus thermophilus S-3 and its application in milk fermentation. *International Journal of Biological Macromolecules, 178*, 263–269. DOI: 10.1016/j.ijbiomac.2021.02.173 PMID: 33639187

Yang, L., Wang, L., Chi, Y., & Chi, Y. (2024). Effect of Whole Egg Liquid on Physicochemical, Quality, Fermentation and Sensory Characteristics of Yogurt. *Foods, 13*(2), 321. DOI: 10.3390/foods13020321 PMID: 38275687

Yang, S., Xu, X., Peng, Q., Ma, L., Qiao, Y., & Shi, B. (2023). Exopolysaccharides from lactic acid bacteria, as an alternative to antibiotics, on regulation of intestinal health and the immune system. *Animal Nutrition, 13*, 78–89. DOI: 10.1016/j.aninu.2023.02.004 PMID: 37025257

Yang, X., Li, A., Li, X., Sun, L., & Guo, Y. (2020). An overview of classifications, properties of food polysaccharides and their links to applications in improving food textures. *Trends in Food Science & Technology, 102*, 1–15. DOI: 10.1016/j.tifs.2020.05.020

Yang, X., Ren, Y., & Li, L. (2022). The relationship between charge intensity and bioactivities/processing characteristics of exopolysaccharides from lactic acid bacteria. *Lebensmittel-Wissenschaft + Technologie, 153*, 112345. DOI: 10.1016/j.lwt.2021.112345

Yildiz, H., & Karatas, N. (2018). Microbial exopolysaccharides: Resources and bioactive properties. *Process Biochemistry (Barking, London, England), 72*, 41–46. DOI: 10.1016/j.procbio.2018.06.009

Yin, X., Li, J., Zhu, L., & Zhang, H. (2024). Advances in the formation mechanism of set-type plant-based yogurt gel: A review. *Critical Reviews in Food Science and Nutrition, 64*(26), 9412–9431. PMID: 37203992

Yin, Z.-y., Yuan, Y., Zhang, R., Gan, J., Yu, L., Qiu, X., Chen, R., & Wang, Q. (2024). Understanding Bacillus response to salt stress: Growth inhibition, enhanced EPS secretion, and molecular adaptation mechanisms. *Process Biochemistry (Barking, London, England), 146*, 412–422. DOI: 10.1016/j.procbio.2024.09.023

Yue, Y., Wang, Y., Yang, R., Liu, D., Cheng, Y., Li, S., Sun, L., Li, B., & Huo, G. (2023). Exopolysaccharides of Bifidobacterium longum subsp. infantis E4 on the immune and anti-inflammatory effects in vitro. *Journal of Functional Foods, 107*, 105699. DOI: 10.1016/j.jff.2023.105699

Zhang, K., Tang, H., Farid, M. S., Xiang, F., & Li, B. (2024). Effect of Lactobacillus helveticus exopolysaccharides molecular weight on yogurt gel properties and its internal mechanism. *International Journal of Biological Macromolecules, 262*, 130006. DOI: 10.1016/j.ijbiomac.2024.130006 PMID: 38331067

Zhang, K., Zhang, T.-T., Guo, R.-R., Ye, Q., Zhao, H.-L., & Huang, X.-H. (2023). The regulation of key flavor of traditional fermented food by microbial metabolism: A review. *Food Chemistry: X, 19*, 100871. DOI: 10.1016/j.fochx.2023.100871 PMID: 37780239

Zhang, T., Liu, W., Lu, H., Cheng, T., Wang, L., Wang, G., & Chen, W. (2023). Lactic acid bacteria in relieving constipation: Mechanism, clinical application, challenge, and opportunity. *Critical Reviews in Food Science and Nutrition*, 1–24. PMID: 37971876

Zhou, Y., Cui, Y., & Qu, X. (2019). Exopolysaccharides of lactic acid bacteria: Structure, bioactivity and associations: A review. *Carbohydrate Polymers, 207*, 317–332. DOI: 10.1016/j.carbpol.2018.11.093 PMID: 30600013

KEY TERMS AND DEFINITIONS

Exopolysaccharides (EPS): These are high molecular mass polysaccharides secreted by microbes mostly LABs, as a result flavor of dairy products is improved.

Fermentation Optimization: Process of upgrading fermentation process. These conditions include temperature, pH, and nutrient availability.

Functional Foods: Foods that provide nutrition and health benefits. These include foods like dairy products enriched with EPS-producing bacteria.

Lactic Acid Bacteria (LAB): Microbes commonly used in fermentation in dairy industry. They convert lactose into lactic acid along with production of side products like extracellular polysaccharides.

Probiotic Properties: Health promoting properties achieved by adding beneficial bacteria.

Sensory Attributes: The attributes which are perceived by five senses. They are enhanced by presence of EPS.

Syneresis Control: Mostly by incorporation of EPS in dairy products like yogurt, syneresis is prevented. This is achieved through water retention and network stabilization.

Viscosity Enhancement: With incorporation of EPS, functional improvement in thickness and viscosity is achieved.

Chapter 4
Integration of IoT in Dairy Processing Facilities

Satyanarayana Sanku
https://orcid.org/0000-0003-0623-9293
Sreenidhi Institute of Science and Technology, India

ABSTRACT

Internet of things technology is found in almost of the all real-life applications. The applications include smart home, smart cities, smart grid, etc. Similar scenarios can be seen in the dairy sector. Agriculture and dairy farms are the backbone of many countries, and their maintenance becomes laborious and expensive. To meet the aforementioned criteria, one needs to depend on the latest technologies such as internet of things, machine learning, artificial Intelligence, blockchain, and supply chain technologies. These technologies services not only provide services to the dairy farms and dairy industries but also provide service to end customers with quality product delivery on time. Therefore, integration of these technologies is inevitable. This chapter provides the details of dairy farms, manufacturing dairy products, and IoT integration in dairy farms and in dairy industries.

INTRODUCTION

Dairy industry is the backbone for many of the countries global economic growth and improved health standards. Most of the under developed or developing countries rely up on traditional methods of yielding the milk and milk products. Such practices although have been used from many decades of years may need the aid from current cutting edge technologies such as Sensor technologies, Internet of Things (IoT) technologies, food processing technologies etc. IoT and their related

DOI: 10.4018/979-8-3693-9949-1.ch004

technologies are game changers in many of the industrial, agricultural, health and many domains including in dairy sector also.

The milk production in cattle farming depends on various factors which include cattle health, cattle food, cattle habitat, cattle living surroundings etc. The Food and Agriculture Organization of United states(FAO) proposes Good Agriculture Practices(GPA) such as Animal Health, Milking Hygiene, Nutrition(feed and water), Animal welfare, Environment and Scio-economic management (Https://Www.Fao .Org/Dairy-Production-Products/Production/Farm-Practices/En/, n.d.).

There have been many sensors developed to monitor all such activities for better yield of the milk from the cattle. The sensors such as smart necklace (a modified fitbit like device) to monitor the cattle health state, a pedometer to monitor the walking steps and distance, an udder sensor to monitor the quality of the milk, an acid monitor sensor inserted in to the cattle rumen to understand the digestive symptoms, a tail monitor sensor used to sense labor pain during pregnancy. These sensors connected together and to internet to form IoT network. Thus IoT technology helps the farmers for smart dairy farming.

Smart dairy farming not just increases the milk quantity but also helps in yield of quality milk. When the quality of the milk is good, that results in producing quality milk products such as curd, cheese, butter, yogurt, lactose free milk, fortified soy milk, ice-cream etc in the dairy industries. The process of producing variety of products from milk needs many procedures to be followed by diary industries. Unique methods and varied methods are followed for producing different diary product. The methods such as fermentation, pasteurization, evaporation, spray drying, standardization, microbial fermentation, filtration, homogenization, deodorization and many more are used for producing varied diary product.

The IoT based systems provide accurate and efficient implementation of various methods for better monitoring and quality outcome of the product. The development of latest technology devices like NIR spectrometers made possible for observing the quality production of various dairy products in dairy industries. Dairy farms and industries produce large wastes, which need to be processed for reusability. The diary waste management also needs monitoring for complete diary industry to be integrated with sensors to make IoT integrated dairy industry.

The International Telecommunication Union for Telecommunications (ITU-T), a global body to set international standards in telecommunication, proposed a IoT reference model (Gunturi et al., 2018), shown in Figure 1, which has four layers and supporting management and security capabilities to ensure that any IoT enabled application has been developed with the necessary requirements. The IoT integration for any application requires the devices, network and protocols used should be mapped to the layers in the model.

Figure 1. IoT reference model (Gunturi et al., 2018)

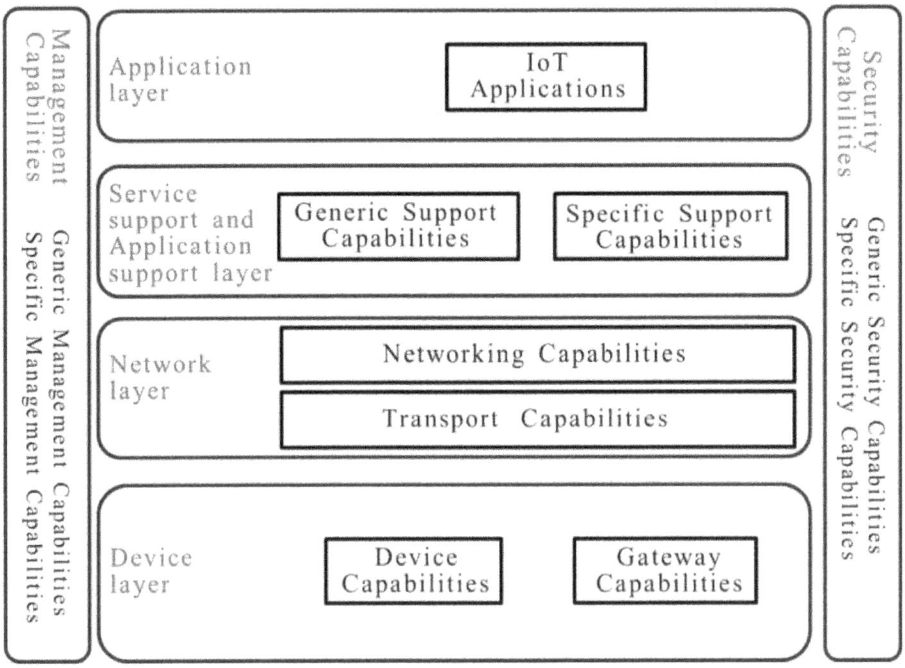

a. **Device Layer:** This layer encompasses physical equipment and sensors used to gather operational data.
   - **Key Elements**:
     - o Sensors for monitoring temperature (storage tanks, pasteurizers).
     - o pH detectors (to assess milk acidity).
     - o Flow meters (for tracking milk volume during processing).
     - o Weight measurement systems (for milk and product quantification).
     - o Identification systems like RFID or barcodes (batch and product tracking).
     - o Robotic milking units (automated milk extraction).
   - **Purpose**:
     - o Captures essential real-time data regarding milk production, quality control, storage parameters, and machinery performance.

b. **Network Layer:** This layer facilitates communication between devices and data processing systems.
   - **Key Elements**:

- Wireless systems (e.g., Wi-Fi, LoRa, ZigBee) to collect data across the farm.
- Wired connections (e.g., Ethernet, fiber optics) for factory-level integration.
- Gateways and edge devices for data aggregation and transfer.

- **Purpose**:
    - Ensures seamless data flow from sensors to local processing units or cloud-based systems, enabling remote oversight and management.

c. **Service Support and Application Support Layer:** This layer provides the computational power and integration needed for data-driven operations.

- **Key Elements**:
    - Analytics platforms (used for evaluating milk quality and process efficiency).
    - AI/ML-based systems (to predict maintenance requirements and ensure quality).
    - Middleware solutions (to connect and manage various IoT devices seamlessly).
- **Purpose**:
    - Converts raw data into actionable insights.
    - Facilitates advanced services such as quality monitoring, yield forecasting, and compliance tracking.

d. **Application Layer:** This layer interacts directly with the end-users through software applications and tools.

- **Key Elements**:
    - Mobile and web apps for monitoring operational data.
    - Alert systems (e.g., notifications for equipment faults or temperature irregularities).
    - Automated process controllers (to adjust processing parameters like pasteurization heat levels).
- **Purpose**:
    - Enables decision-makers to access real-time insights, control processes, and ensure quality standards are met.

e. **Management Compatibility:** This layer deals with managing the operational lifecycle and performance of the IoT infrastructure.

- **Key Elements**:
    - Tools for managing devices (e.g., updates, configuration, troubleshooting).
    - Workflow management platforms (to enhance production and resource use).
    - Systems for tracking utilities like energy and water consumption.

- **Purpose**:
  - o Oversees system efficiency, resolves hardware/software issues, and ensures optimal resource utilization.

f. **Security Compatibility:** This layer protects the IoT ecosystem from risks and ensures data privacy.

- **Key Elements**:
  - o Encryption methods (to secure data storage and transmission).
  - o Authentication mechanisms (to restrict unauthorized access).
  - o Intrusion detection systems (to identify and prevent cyber threats).
- **Purpose**:
  - o Safeguards sensitive operational data and prevents unauthorized access to crucial systems.

Table 1. IoT Layers and their roles in dairy processing

Layer	Primary Role in Dairy Processing
Device Layer	Collects key data on milk production, quality, and equipment.
Network Layer	Transfers data efficiently for further analysis.
Service Support Layer	Analyzes data and supports decision-making processes.
Application Layer	Offers interfaces for monitoring and controlling operations.
Management Compatibility	Manages devices, workflows, and resource allocation.
Security Compatibility	Protects data integrity and ensures system resilience.

This revised mapping supports a well-integrated, efficient, and secure IoT ecosystem for smart dairy management.

The hybrid framework of integrating latest cutting edge technologies such as IoT, AI, ML, edge computing, robotics, drones, and cloud collaboratively along with IoT for an intelligent dairy farm was demonstrated by the authors in (Gehlot et al., 2022) with the help of the Figure 2. The end nodes in the sensing layer gather the real time raw data from the various sensors employed for what it is defined for. These devices send the data through serial/parallel communication protocols to the edge devices at the edge layer, where the certain level of computations happen and then forwarded to the gateway layer by using link layer protocols such as Zigbee, LoRa, Bluetooth, Wi-fi, etc. The drones and robotic devices also help the collection of raw data from the environment as well as monitoring the significant assignment aiding the edge layer for making processing easy. The gateway devices offer the data to get forwarded to the cloud layer with the help of internet protocols such as TCP/I or UDP. The cloud layer offers various services such as Infrastructure as a Service (IaaS), Platform as a Service (PaaS) and Software as a Service (SaaS) to the customers on rental basis.

Figure 2. Hybrid framework with the integration of multiple emerging technologies (Gehlot et al., 2022)

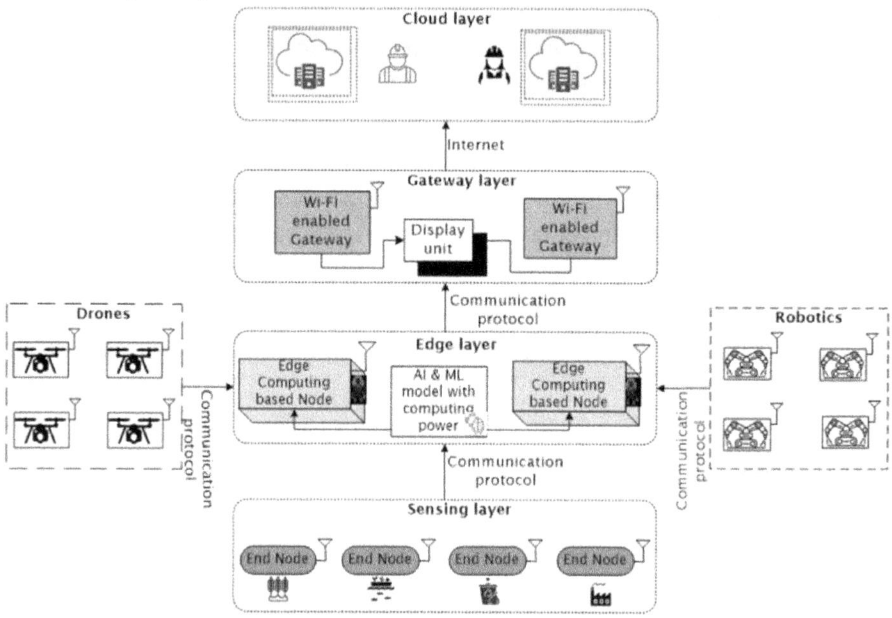

LITERATURE REVIEW

Research works were carried out for improvising afore mentioned methods by many researcher. The following observations were made:

a) Nuclear Magnetic Resonance (NMR) relaxometry and high-field Magnetic Resonance Imaging MRI techniques help categorizing/ comparing traditionally made Italian cheese called Fiore Sardo (FS) cheese i.e from raw milk with that of the cheeses made from heat-treated milk. These methods evaluate the quality of cheese produced can integrate with IoT technology. (Anedda et al., 2021)
b) The traditional electrochemical technology which is the most common method for monitoring the fermentation process and determining the pH of fermented dairy products can be replaced by an electronic nose, which classify different bacteria groups and time points and also monitors the formation of the aromatic compounds during the fermentation process. Integrating such devices to IoT network enhances the accuracy and save time. (Aydogdu et al., 2023)

c) Many sensors and sensor technologies are developed to enhance quality controls in the food chain process in diary processing industries. (Fernández González et al., 2024)
d) A weight sensor measures the milk vessel weight and the sensing system checks the quantity of the milk. And an RFID reader connected with the milk monitoring mote identifies the milk vessel through the RFID tag attached to it. This work was extended for n numbered milk monitoring motes in a diary farm, forms a block chain process, when the system is connected to the server. To connect to the cloud server, it should establish communication with a gateway, so the LoRa module embedded in the milk monitoring mote empowers it to connect to the gateway. The LoRa module in the gateway serves as a receiver unit, and the Wi-Fi module allows it to connect to the internet and log data to a cloud server. (Gehlot et al., 2022)
e) An electronic nose (E-nose) consisting of seven metal oxide semiconductor sensors is developed to identify milk sources (dairy farms) and to estimate the content of milk fat and protein which are the indicators of milk quality. For milk source identification, the features based on milk odor features from E-nose, composition features (Dairy Herd Improvement, DHI analytical data) from DHI analysis and fusion features are analyzed. (Mu et al., 2020)
f) Yogurt manufacturing main depends on proper milk fermentation. Control of the milk fermentation process end point in industrial yogurt production is very important because of its relationship with product quality. If the pH is high, fermentation is incomplete, resulting in yogurt with an undesirable taste and texture. If the pH is too low, the yogurt may become too sour, and the texture may separate and release whey. Also, yogurt's final pH in industrial production plants requires values lower than 4.6 due to food safety reasons. All these required inline and nondestructive methods for monitoring milk coagulation, for which IoT based system integration in the dairy process is beneficial. (Salah et al., 2021)
g) Waste management in the dairy industries for the eco-friendly and green environment. (Botondi et al., 2023)

IOT DEVICES IN DAIRY FARMING

Managing the health and production of the cattle in the dairy farms maintained by the dairy farmers becomes incredibly difficult when the herds of cattle increase in count. The solution to this problem is mostly given by IoT technologies. Monitoring of various parameters related to the milk yielding animal is essential to produce the milk for global need, because milk is the main contributing food item for calcium

(49% of global nutrient availability), Vitamin B2 (24%), lysine (18%), and dietary fat (15%), and contributes more than 10% of global nutrient availability for a further five indispensable amino acids, protein, vitamins A, B5, and B12, phosphorous, and potassium (Smith et al., 2022).

The IoT technologies at the ground level use various sensors for monitoring the activity of cattle in the dairy farms, one such implementation is a Precision Livestock Farming (PLF), which is based on the use of sensors to monitor individual animals in real time, enabling farmers to manage their herds more efficiently and optimise their performance. The integration of sensors and devices used in PLF with the Internet of Things (IoT) technologies (edge computing, cloud computing, and machine learning) creates a network of connected objects that improve the management of individual animals through data-driven decision-making processes (Tangorra et al., 2024).

Cow Details Collection/Monitoring

For sensing and monitoring the parameters related to the cattle in the dairy farms, they need to be connected to different sensors at different places. The implantation of sensors on to the cattle is shown in Figure 3, which makes the cattle an innovative one. The cow equipped with Tail sensor, Ear sensor, Reticulorumen sensor, leg sensor and noseband sensor sense and collect the detail raw information of the cow.

Figure 3. Innovative cow in precision livestock farming

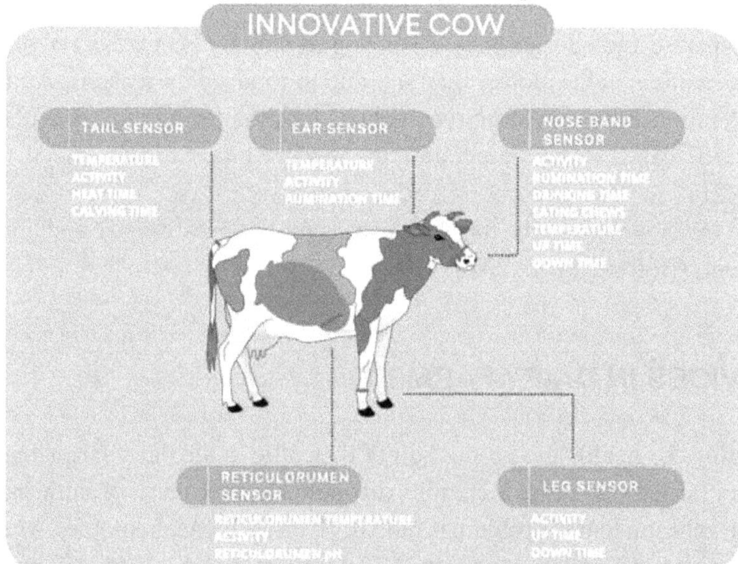

Table 2 below shows the details of all kinds of sensors used for mounting on the cattle and in the dairy farms.

Table 2. Wearable sensors on cattle for IoT based precision livestock farming

Purpose	Sensor	Devices	Function	Disease/ Problem	Remarks/ other info
Health	Temperature sensor	1. Rectal thermometers(inserted inside), 2. Infrared temperature sensors (placed near eye or ear or on udder), and **3. I**mplantable or wearable sensors(attached to leg, ear or placed inside rumen)	Monitors body temperature	It helps to detect Fever, ovarian cysts, pneumonia, retained placenta, mastitis	Good health (38.0°C and 39.3°C (100.4°F to 102.7°F)).
	Acid monitor sensor	1.Bolus Sensors(orally administered and resides in the rumen). 2.Rumen Catheter Systems (Temporarily inserted). 3.Wearable Sensors with Indirect pH Estimation. Typical devices: SmaXtec Bolus, eCow Bolus.	measures the pH level in the rumen thus monitors the digestive symptoms	It helps prevent conditions like **ruminal acidosis**.	6.0 to 7.0 in cows fed a forage-based diet. Slightly lower (5.5 to 6.5) in cows fed a high-grain diet (common in dairy farms). Acidosis (pH < 5.5). Alkalosis (pH > 7.0)
	Tail monitor sensor	1. Wearable Tail Sensors(Attached to the tail) 2. Multifunctional Tail Sensors Attached to the tail)	To sense labor pain during pregnancy.	It helps us to know Estrus (Heat), Calving, Irritation at the back.	
	Milk color sensor	Color sensor	Cow with Mastitis(reddish, brownish, or yellowish milk)	It helps to detect Mastitis	Normal Milk(White or slightly creamy)

continued on following page

Table 2. Continued

Purpose	Sensor	Devices	Function	Disease/ Problem	Remarks/ other info
Activity / location	Pedometer	AfiAct II Leg Pedometer AfiTag RumiWatch	To monitor walking steps and distance	Activeness of the cow	
	GPS/GNSS	GPS Ear Tag	To locate the cattle position in the field.	Location identification	Grazing, feeding, lying, behavior and welfare monitoring
	Radio-frequency identification	ISO standard UHF 915 MHz RFID cattle tags	To identify the individual cattle among many.	Count	
Milk production	udder sensor	udder sensor	To monitor the quality of the milk	It helps to detect Mastitis	

Environment Monitoring in the Dairy Farm

The Table 3 below shows the details of these sensors.

Table 3. Environment monitoring sensors for IoT based precision livestock farming

Purpose	Sensor	Function	sensor placement	Suitable value
Indoor	Temperature sensor	To sense the temperature inside the cattle shed	Indoor unit	−0.5 to 20.0 °C.
	Humidity sensor	To sense the humidity in the cattle shed.	Indoor unit	40-80%
	Air Quality sensor	To sense air quality in the shed.	Indoor unit	below 100
Outdoor	Temperature sensor	To sense the temperature inside the cattle shed	Outdoor unit	Depends on the cattle and its breed
	Humidity sensor	To sense the humidity in the cattle shed.	Outdoor unit	Depends on the cattle and its breed
	Air Quality sensor	To sense air quality in the shed.	Outdoor unit	Depends on the cattle and its breed
	Rainfall sensor	To sense the rainfall while grazing	Outdoor unit	Depends on the cattle and its breed

The comparison of cutting edge technology live stock management with that of the old traditional live stock management is illustrated with the help of Figure 4.

Figure 4. Tradition live stock management vs. cutting edge technology live stock management (Neethirajan, 2023)

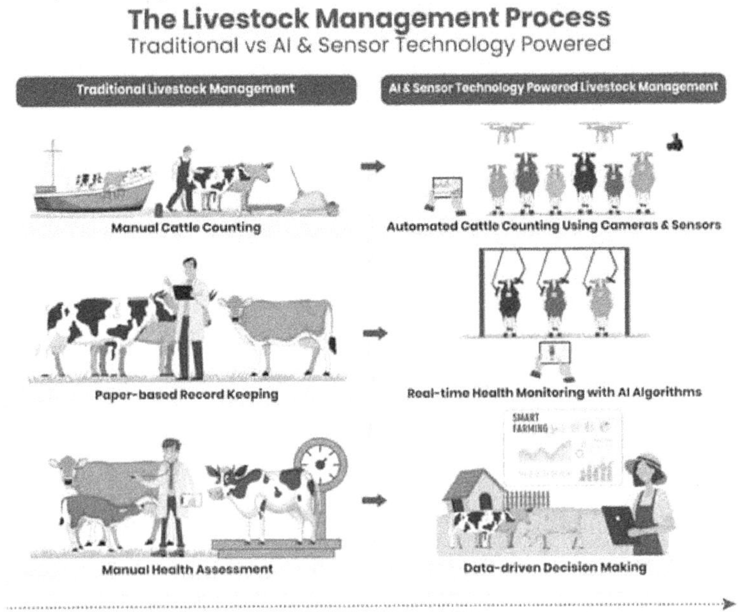

IoT Enabled Dairy Farm

Many researchers worked on attachment of sensors to cattle and collecting data from them. IoT based livestock mentoring eases the dairy farm maintenance by reducing labor cost and safeguards the cattle. The three different architectures are proposed by the authors, which is shown in Figure 5; One being the architecture describing the connectivity of end devices mounted on the individual cattle in the farm through gateway and TCP/IP protocol to the computer system. The End devices collect the raw data such as location of the cattle in the field and send the data through LoRa LPWAN technology. The End Device (ED) has an Adafruit Feather M0 RFM95 microcontroller with embedded LoRa modem, a LiPo rechargeable battery, and an EEPROM memory. Along with the aforementioned hardware, the sensors such as GNSS Neo-6m, orientation acceleration MPU-6050 and DS18B20 Body temperature sensor are also used to monitor and collect the details about the cattle (Germani et al., 2019; Hu et al., 2023).

The Second proposed architecture demonstrated the monitoring of indoor facility of the cattle: monitoring temperature and humidity by AM2321 sensor, monitoring illumination by GY-2567 sensor, Co2 concentration monitoring with SEN0219 sensor and NH3 concentration monitoring with M2137 sensor. All these sensors are controlled by an arduino Mega board, which is finally connected to the Raspberry Pi board. The third proposed architecture has a Raspberry Pi board connected with the AM2321 sensor for monitoring outdoor weather conditions and INA219 current and voltage monitoring from the solar panel charging the battery.

Figure 5. IoT based livestock monitoring facility (Germani et al., 2019)

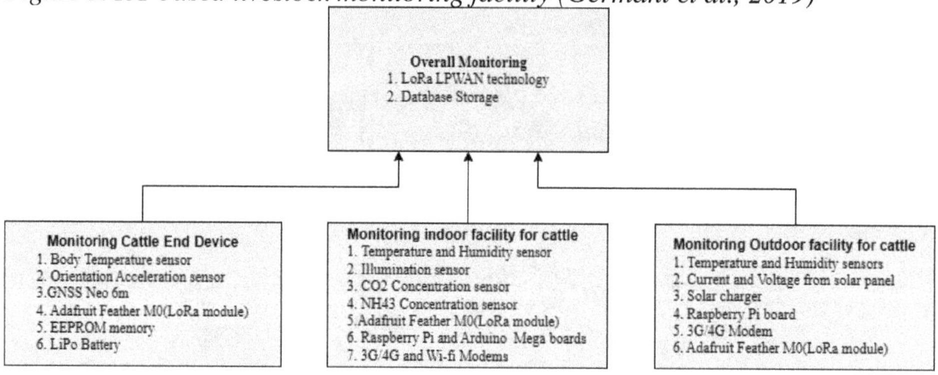

The basic structure of smart dairy farm with required number of sensors used to monitor the activities in the dairy farm are discussed by the authors in (Hu et al., 2023) is shown in Figure 6. In this two scenarios are considered, one being the cow shed and another one is the milking parlor. The cows are monitored by devices such as drones during feeding outdoor, automatic stacking for proper arrangement of cows in the shed and ear tag composing of sensors to measure height, weight, body length and genome of the cows. The cows shed are monitored by gas detection sensors like: NH3 sensor, CO2 sensor and CH4 sensor and the environment sensors measure temperature and humidity conditions in the cow shed and mask sensor for air quality. The milking parlor has intelligent weighing system for individual cow milk.

Figure 6. Facility in smart dairy farm (Hu et al., 2023)

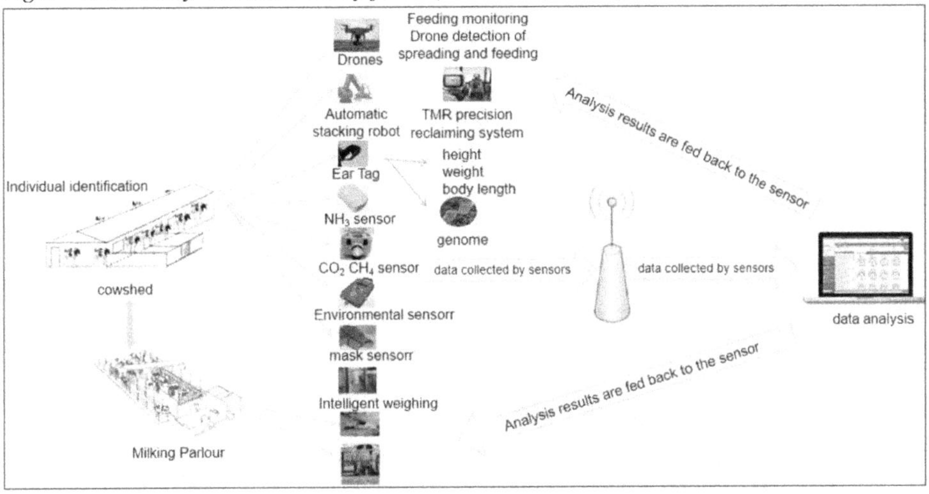

IoT Integration at Dairy Farms

Milk giving cows are maintained in separate farms where the micro and macro weather conditions in the farm are recorded along with the health data with the sensors attached to the cows. These data are then stored in the AWS cloud platforms which helps the farm keepers to know the status of the cow and the milk production. All this setup is shown in the Figure 7 (Dineva & Atanasova, 2023).

Figure 7. Cows at the farms

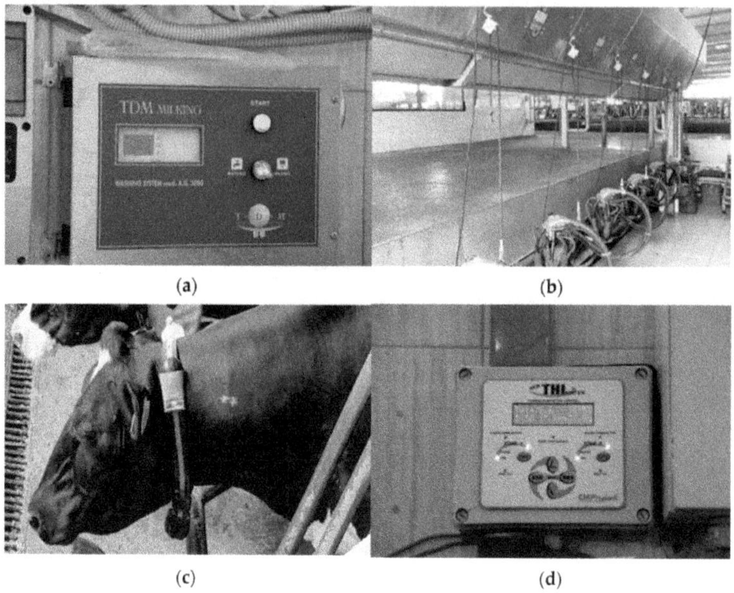

Figure 8. Integration of latest technologies in a milk production environment (Dineva & Atanasova, 2023)

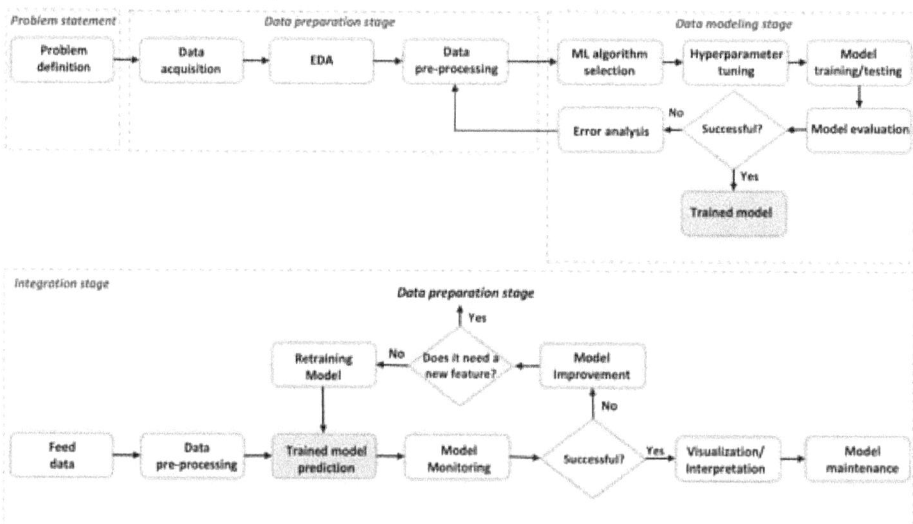

The integration of IoT and other enabled technologies in to the dairy farms, which is shown in Figure 8, need to follow initial and sequential steps: The procedure starts from the problem statement, which is formulated to classify the cows based on its health condition into three categories healthy, unhealthy and suspicious. Data preprocessing is performed later, after data collection from various sensors attached to the cows and from local weather condition in the farm and exploratory data analysis is performed on the collected data to validate the data for different purposes like feature selection, model selection, and hyper-parameter tuning. As the part of data modeling next, suitable machine learning model is selected, hyper-parameter is tuned and then model is trained tested and evaluated.

Now for IoT integration, data is fed, pre-processing is performed, model is monitored and validated and results are visualized and interpreted. The performance of the model was evaluated using parameters such as accuracy, precision, recall and F1 score.

AWS cloud platform can be chosen as a service provider for IoT data integration, where storage, preprocessing and implementation of machine learning model are possible. This is shown in Figure 9 below.

Figure 9. AWS as a service provider for IoT integration

IOT IN MILK COLLECTION AND QUALITY ASSURANCE

Milk is the main source of all the dairy products. Milk is collected from different cattle like cows, goats, sheep and camel. Among which more percent of the milk contributor is cow. Milk is the most consumed food product all over the globe. The statistics show that there will be an increase of 1.6 percent World milk production annually between 2020 and 2029 and can reach 997 million tons in 2029(*Global Dairy Industry and Trends - Feed & Additive Magazine*, n.d.). India at the top of the list in milk production as well as in consumption with 81 million tons in 2021 has more responsibility along with other countries to ensure quality milk production. To ensure the health of the global population, maintaining the milk and dairy products quality are deliberately essential.

The collected milk quality may be altered unknowingly because of bad farm practices; i.e milk may have contaminations. Or deliberately may get added with some chemicals for greed of earning more money from the market; i.e. milk undergo adulteration.

Milk Contamination

The unhygienic conditions at the farms during milk extraction and storage cause contaminations by pathogens. The tradition methods of heating the milk for the removal of the pathogens cause reduce the nutritional content and change the structure of fat and proteins. So, the acoustic cavitation is performed on the milk by using ultrasound waves causing the pathogens to destroy in the milk. There are two types of acoustic cavitations: a) High frequency ultrasound, in which the time period of the wave is less under constant pressure conditions, so the cavitations happen quickly. b) Low frequency ultrasound, in which the time period of the wave is more as compared with earlier one under constant pressure conditions, so the cavitations happen lately. Figure 10 shows the acoustic cavitations process.

Figure 10. Acoustic cavitations on milk (Zadeike & Degutyte, 2023)

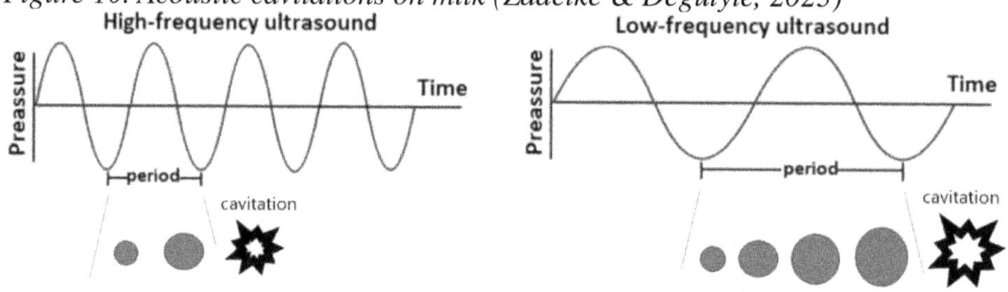

Milk Adulteration

Due to the increase in the milk consumption requirements and demand-supply requirements, there are many possibilities of technological frauds for milk and dairy products. The possible adulterations of milk are shown in Figure 11.

Figure 11. Types of milk adulterations (Ionescu et al., 2023)

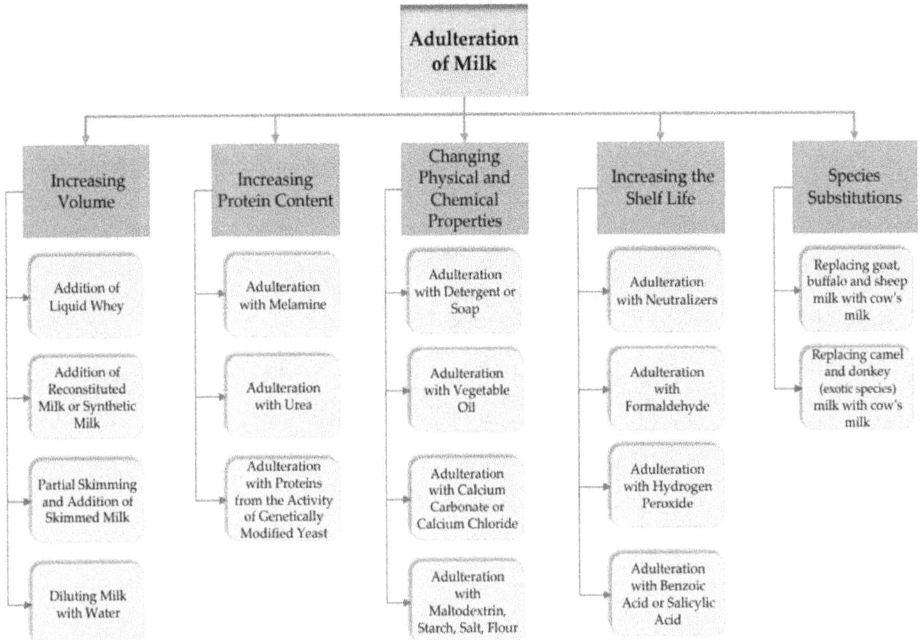

Adulteration of milk and dairy products are to be eradicated with the aid of latest technologies. There are basically three methods identified a) physical methods, which depends on the physical properties of the milk like freezing point, density, refractive index etc. b) chemical methods, which are based on the pysico-chemical changes in the milk, example color based reaction observed by colorimetric detection. c) Instrumental methods, which are the modern biotechnological methods and provide accurate and rapid results such as molecular fluorescence, gas chromatography, high performance liquid chromatography, infrared spectrometry, measurement of refractive index, scanometry and UV-vis spectrophotometry(Ionescu et al., 2023).

For quality assurance of the milk, the parameters used are pH value, viscosity, color, fat content etc.

Table 4. Parameters for milk quality assurance

Parameter	Sensors	Typical values	Devices in market
PH	pH sensor	6.2 to 6.8	Orion Star™ A111 (Thermo Fischer Scientific Inc., Waltham, MA, USA) pH meter
Color	Colorimeter	Cow's milk is yellowish white. Buffalo milk is greenish white.	1. CR400, Konica Minolta, Tokyo, Japan. 2.MiniScan XE Plus portable handheld color spectrophotometer (HunterLab, Reston, VA, USA)
Fat content	Butyrometer (**Gerber method**)	3.25 to 3.5% (for cow milk and varies by breed of the cow) • Toned milk (3%), Double toned milk (1.5%), • Skimmed milk (> 0.5%), • Standardized milk (4.5%), • Full cream milk (6%), • Recombined milk(3%)	• Many are available
temperature	Temperature sensor	4–5°C	Many are available
Tank level	Ultrasonic sensor	Depends on tank capacity	Many are available

One of the latest technology sensors used for identifying the fraudulent in milk production is Electronic nose (Balivo et al., 2023). This system uses an electronic nose, which identifies the quality milk by discrimination between the milk extracted from grazing cows with that of milk extracted from confined feeding system. The MOS sensor captures the odor details from the milk samples and then data is classified. The pattern recognition techniques such as Linear Discriminant Analysis (LDA) is performed over the data vectors generated by odor signal processing to recognize the quality milk.

VARIOUS DAIRY PRODUCTS AND INDUSTRIAL NEEDS FOR THEIR PRODUCTION

The collected raw milk from various dairy farms are treated and preprocessed. There are thermal and non thermal methods of processing the milk. The thermal process includes: Thermization, Pasteurization and Sterilization, Ohmic Heating (OH). And the non-thermal methods include: Pulsed Electric Field, Supercritical Fluid Technology, UV Radiation, Ionizing Radiation (e-Beam, Gamma), Cold Plasma—Nonthermal Plasma (NTP), High-Pressure Processing (HPP), Ultrasounds (USs), Membrane Technologies. These methods are generally performed to keep the milk microbiologically safe and to reduce the enzyme activity, so as to retain its quality, durability and make it ready for production of dairy products such as Curd, Cheese, Butter, Ghee, yogurt, etc. (Żbik et al., 2024).

Some of the most frequent methods used for milk processing are:

a) Pasteurization: Pasteurization is a mild heat treatment done to the milk i.e. heating milk to such temperature and for some periods of time, so as destroy any pathogens which may be present, and causes minimum changes in the composition, flavor and nutritive value. This process is performed for two main reasons, one being to destroy bacteria from the milk and second being to eliminate spoilage bacteria to maintain milk quality. The milk sterilization methods are followed worldwide are
    i) pasteurization (heating to 72–75°C for the duration of 15–16 s or heating to 80–85 °C for the duration of 10–15 s) and
    ii) Ultrahigh Temperature (UHT) sterilization (heating to 135–150°C for the duration of 4–15s)
b) Homogenization (Shao et al., 2023): It is a process of exposing the milk to the pressure for certain duration of time so as to break down the fat globules present in it and to maintain the organoleptic quality and stability of pasteurized milk. The authors have evaluated the best storage time, duration of time during which the milk is exposed to the homogenized pressures. The three storage durations, three homogenization durations and homogenization pressure were considered. The results were evaluated, one with the PEN2 portable electronic sensory nose which have 10 metal oxide sensors to collect the bionic olfactory information and for evaluating the physical properties of the milk, subjective analysis were performed, in which the different storage time, homogenization durations and pressures samples were given to trained subjects to provide the physical properties of the milk like taste, smell texture etc. Results of their work are summarized in Table 5.

Table 5. Results of homogenization process under different considerations

Storage time	**Homogenization duration**	**Homogenization pressure**	**Evaluation**
8 Hours, 72Hours and 168 Hours	40s, 56s and 60s	0bar, 150bar, 300bar	Electronic nose technology and Physical properties
Outcome			
Input	**Storage time**	**Homogenized pressure**	**Results**

Table 5. Continued

Storage time	Homogenization duration	Homogenization pressure	Evaluation
The pasteurized milk sample	8 h	Under the three homogenization pressures	A good fresh milk flavor and its strong fat flavor affected the taste.
The pasteurized milk sample	72 h	Under 150-bar homogenization pressure	Uniform texture, excellent milk flavor, and delicious taste, and the stability of milk reached the best state
The pasteurized milk samples	168 h	Under three homogenization pressures	Increased metallic, sour, and salty tastes in varying degrees, affecting the flavor quality and stability of milk.

Processing of milk in certain way produces various dairy products, which can be seen in Figure 12.

Figure 12. Dairy products from milk

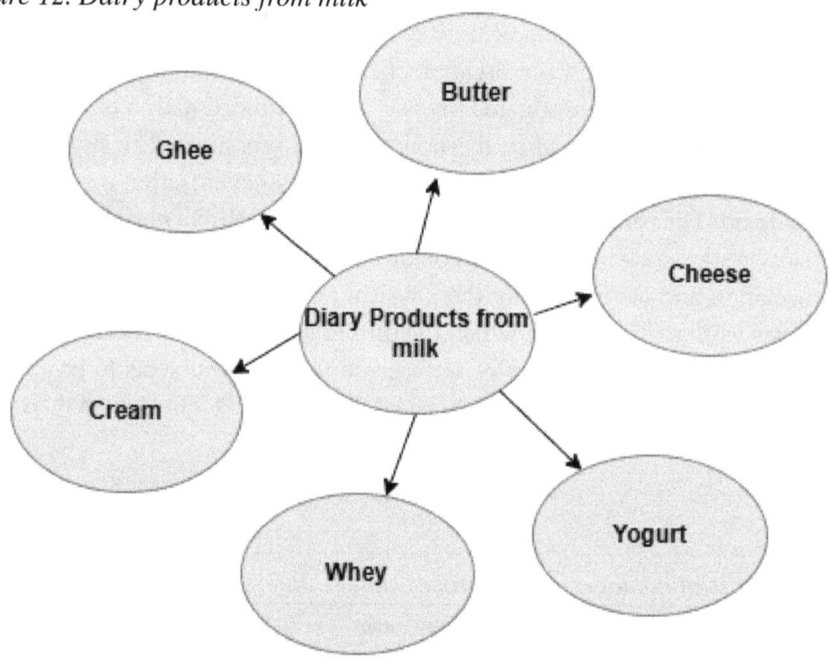

PRODUCTION OF CHEESE

Cheese is one among the milk products, which is used in tasty food making recipes and are of different types. Cheese can be stored for long duration and has many health benefits.

Cheese Manufacturing

The process of manufacturing cheese from the milk is shown in the Figure 12 (Scudino et al., 2023).

Figure 13. Cheese preparation process

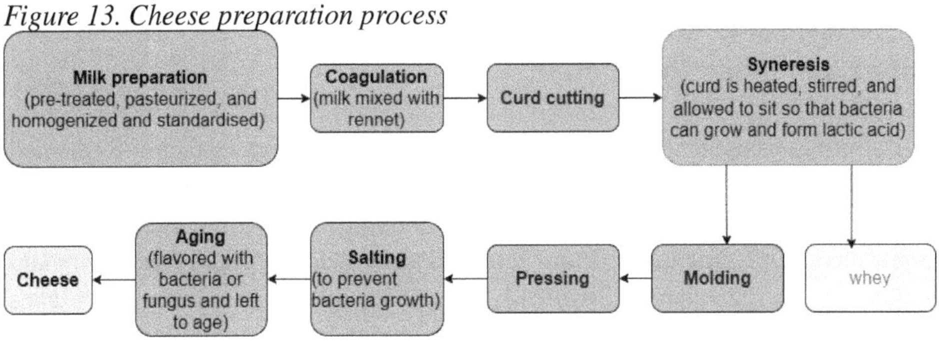

Cheese Quality Evaluation

Different methods of evaluating the cheese quality are tabulated in Table 6 (Bosakova-Ardenska, 2024).

Table 6. Methods for evaluating cheese quality

Different Methods
Ultrasound signal
Near Infrared (NIR) signal based
Magnetic Resonance Imaging based
Digital image based /computer vision based.

Ultrasound signal (Scudino et al., 2023): Ultrasound processing has been established to be a safe and environmentally friendly alternative to conventional heat-based processing technologies that lead to the degradation of milk quality.

Ultrasound signals have a wide range of applications in the dairy industry in two main directions—firstly, as a tool for the microbiological enhancement of milk during different processes of its transformations (high-intensity ultrasound signals), and secondly, as a tool for the quality assessment of different milk products (low-intensity ultrasound signals)

NIR Spectrometer (Pu et al., 2021): NIR spectrometer is used to measure major things in milk like: moisture, fat, protein, lactose. And minor constituents like salt, pH, water-soluble primary amines) in different dairy products which included liquid milk, milk powders, casein and caseinates, butter and cheese.

PRODUCTION OF BUTTER AND GHEE

Butter and Ghee Production

Butter and Ghee are the edible products formed from fermented or non-fermented milk of cattle like cow, goat and buffalo (yak in some parts of the world). As ghee is rich in saturated and unsaturated fatty acids composed of 98.9% fat, its consumption is prevalent. Different methods are followed to produce butter, Tibetan way and Indian way are illustrated in Figure 14. The production of Ghee from butter takes few more steps, (Chi et al., 2024).

Figure 14. Process flow for butter and ghee production (Chi et al., 2024)

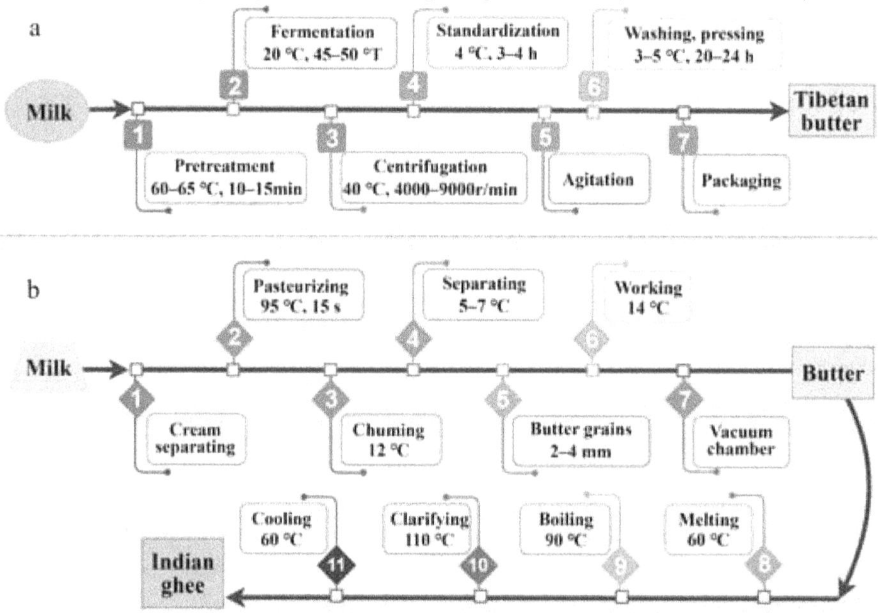

Detecting Ghee Adulteration

There are various methods used for detecting the adulteration done to the ghee and are highlighted in Table 7.

Table 7. Table methods for detecting adulteration in ghee

Methods	Types
Spectroscopic methods	Fluorescence spectroscopy
	Raman Spectroscopy
	ATR-FTIR Spectroscopy
Physico-chemical parameters	Reichert-Meisel value
	BR reading
	Polenske value
	RM value
Chromatographic	Gas Chromatographic
	High performance liquid Chromatographic
Some engineering techniques	Electronic nose
	Computer vision approach
	LF-NMR
Others	Ultrasonic velocity
	DPPH impregnated paper
	Digital PCR Assay

PRODUCTION OF YOGURT

Yogurt is a fermented dairy product recognized for its mineral content, including potassium, zinc, phosphorus, and calcium. Calcium, in particular, plays a crucial role in the development of bones and teeth, supporting children's growth and reducing the likelihood of osteoporosis in adults [2,3]. Additionally, regular yogurt consumption offers several health benefits. These include improved digestion of proteins and sugars compared to milk, enhanced peristalsis due to lactic acid content, and easier digestion. It also supports oral health, promotes the colonization of beneficial microorganisms in the gastrointestinal tract, strengthens the body's support systems, reduces inflammation, boosts immune function, aids in the production of hormones and enzymes, and facilitates better absorption of minerals [4].

Frozen Yogurt Production Process

1. **Milk Preparation:** Pasteurized whole milk is heated to a temperature of 70 °C ± 2 °C.
2. **Incorporating Ingredients:** Specific quantities of sugar, maltodextrin, Grade A nonfat dry milk, corn starch, and hesperidin are added based on the formulation. The ingredients are thoroughly blended using stainless steel stirring rods until fully mixed.
3. **Pasteurization:** The mixture is pasteurized at 75 ± 1 °C for 30 minutes, and then cooled to 43 ± 1 °C.
4. **Fermentation:** The cooled mixture is inoculated with a 1:1 ratio of *Lactobacillus bulgaricus* (LB-12) and *Streptococcus thermophilus* (ST-06). Fermentation occurs at 43 ± 1 °C in an incubator until the pH level reaches 4.7 ± 0.1.
5. **Aging:** The yogurt is placed in storage at 4 ± 1 °C for 24 hours to allow aging.
6. **Mixing and Freezing:** After aging, the coagulated mixture is stirred manually for 60 seconds to eliminate clumps. The smooth mixture is then transferred to a batch freezer and frozen for 9 minutes and 30 seconds.
7. **Packaging and Storage:** The frozen yogurt is packed into polypropylene containers of various capacities (59, 163, 355, 473, and 946 mL), sealed with lids, and stored at −25 ± 2 °C for testing at intervals of 0, 30, and 60 days (Cedillos et al., 2024).

WASTE MANAGEMENT IN DAIRY INDUSTRIES

The production of different dairy products leads to generate dairy wastes, which is needed to be treated before released into the environment. Therefore dairy waste management most significant and is efficiently possible with IoT integration.

Waste Water Management Using IoT

Water being precious natural resources need to be treated or made used for useful purposes, after being used in the dairy farms for washing cattle sheds and cattle. The authors in their work has proposed a waste water treatment system using IoT and experimented in the pig farm, where the water used for washing and cleaning pig and their farms. The IoT based water quality sensors are used to find the removal efficiencies of biochemical oxygen demand (BOD), chemical oxygen demand (COD), and suspended solids (SS) from waste water in the pig farms. The sensors used by them are BOD sensor, COD sensor, SS sensor Electrical conductivity (EC) sensor and pH sensor. The data from these sensors are integrated together and transmitted using 4G network for remote monitoring with the aid of the AQWEB app. Their evaluated results have proved that using their system water was free from the radicals mentioned above with considerable percentages (Su et al., 2020).

Dairy Waste Management

Cattle in the dairy farms produce large quantity of dung, which is the dairy waste. This dairy waste is processed to produce biogas. The biogas is the source for renewable energy source for many purposes such as domestic cooking fuel, driving source for electric pumps in agriculture and for transportation. The authors in their article have reviewed the significance of small-scale biogas technology and its relevance in rural environment of India. An IoT based biogas monitoring system can be installed with low cost which monitors the efficient production of biogas in the plant. The authors have proposed a conversion of bio gas plat into a IoT enabled biogas system with various necessary sensors (Pandey et al., 2021; Kalamaras et al., 2024).

IOT INTEGRATION INTO VARIOUS SECTORS OF DAIRY PROCESSING

The integration of various sensors together, through cattle health in dairy farms to the finished dairy product, is essential for efficient utilization of cutting edge technologies:

The integration of IoT sensors in various steps of dairy processing is shown in Figure 15.

a) Dairy Farms: At dairy farms, as discussed in section 3, the sensors are to be placed for cattle health and habitat monitoring, Indoor and Outdoor farm environment monitoring and Milk quality sensing. These sensors collect the raw details from the dairy farms and forward the data to the cloud for necessary processing. Thus the integration of sensors at the dairy farms is layer-1, which is called as Dairy farm layer
b) Milk Collection Centre: Milk from dairy farms is collected at the bulk level so as to supply it to the dairy processing industries. The quantity sensors measure the weight of the collected milk and the requirement for supplying to dairy product manufacturing industries. The quality sensors ensure the qualities as per the norms are achieved or not. The adulteration sensors figure out the milk with adulteration. The storage sensors are used during pre-processing of milk for various dairy products manufacturing and for storage. The layer-2 called

Milk collection layer performs the sensing related to milk collection and quality assurance. The section 4 discussed the details of this layer.

c) Curd and Butter milk Production Unit: Quality and processing sensors collect the data and forward to the cloud where product durability and taste are estimated.
d) Cheese manufacturing Unit: There are varieties of cheese produced; every type has its own taste and flavor. The IoT devices aid in production of cheese and packaging of it. Section 6 provides the details about the cheese production and adulteration detection in it.
e) Butter and Ghee Production Unit: The integration of sensors during manufacturing of butter and ghee aids the quality production and packing of this dairy product. The section 7 provides the details of making and adulteration detection methods for which sensors are implanted in the line of process.
f) Yogurt and Cream Production Unit: Their production line contains required number of sensors for quality and packing. Integration of these sensors makes the monitoring process easy. Section 8 explained the required conditions for yogurt making.

Figure 15. Integration of IoT device into dairy processing

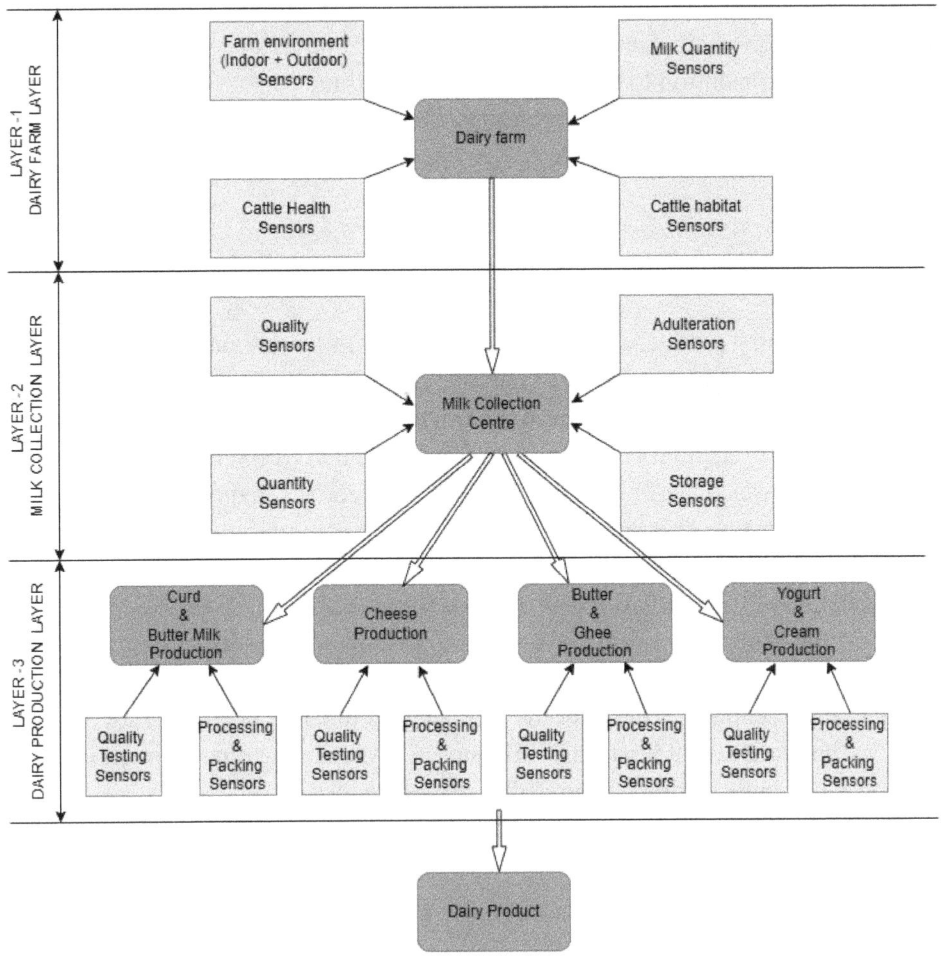

FUTURE PROSPECTIVE DIARY PROCESSING

The integration of IoT devices are very much essential in some of the most requires areas in the processing of milk and dairy products. Such as

1. Integration of Big data analytics with Machine Learning and Artificial Intelligence algorithms for cattle management and dairy farm maintenance.
2. Use of Artificial Intelligence enabled Robotics for milk extraction by monitoring the health conditions of individual cattle in the farms.

3. Implementation of Artificial Intelligence and Machine Learning algorithms in dairy industries ensures the quality, safety and productivity of diary products.
4. Use of Supply chain management for production of quality dairy products from milk and distribution to the dairy parlors for selling.
5. Use of block chain management ensures the safety, transparency and traceability of the dairy products.

CONCLUSION

Milk and dairy products play a significant role in human nutrition, providing essential nutrients such as proteins, vitamins, and minerals. These foods are particularly noted for their high calcium content, which is vital for bone health, growth, and maintenance. Regular consumption of milk and dairy products has been associated with a lower risk of conditions such as osteoporosis, hypertension, and cardiovascular diseases. They are also a valuable source of vitamin D, riboflavin, and potassium, contributing to overall dietary quality and health outcomes. Furthermore, fermented dairy products like yogurt and cheese offer additional benefits, including improved digestion and a positive impact on gut microbiota (Haug et al., 2007). So, the quality in extraction of milk and quality in production of associated dairy products are essentially needed for the wellbeing of the ones health and for healthy society. Therefore the integration of IoT into every step in the process starting from cattle sheds to end customer ensures the mentioned task.

REFERENCES

Anedda, R., Melis, R., & Curti, E. (2021). Quality Control in Fiore Sardo PDO Cheese: Detection of Heat Treatment Application and Production Chain by MRI Relaxometry and Image Analysis. *Dairy*, *2*(2), 270–287. DOI: 10.3390/dairy2020023

Aydogdu, T., O'Mahony, J. A., & McCarthy, N. A. (2023). pH, the Fundamentals for Milk and Dairy Processing: A Review. *Dairy*, *4*(3), 395–409. DOI: 10.3390/dairy4030026

Balivo, A., Cipolletta, S., Tudisco, R., Iommelli, P., Sacchi, R., & Genovese, A. (2023). Electronic nose analysis to detect milk obtained from pasture-raised goats. *Applied Sciences (Basel, Switzerland)*, *13*(2), 861. DOI: 10.3390/app13020861

Bosakova-Ardenska, A. (2024). Recent Trends in Computer Vision for Cheese Quality Evaluation. *Engineering Proceedings*, *60*(1), 12.

Botondi, R., Lembo, M., Carboni, C., & Eramo, V. (2023). The use of ozone technology: An eco–friendly method for the sanitization of the dairy supply chain. *Foods*, *12*(5), 987. DOI: 10.3390/foods12050987 PMID: 36900504

Cedillos, R., Aleman, R. S., Page, R., Olson, D. W., Boeneke, C., Prinyawiwatkul, W., & Aryana, K. (2024). Influence of Hesperidin on the Physico-Chemical, Microbiological and Sensory Characteristics of Frozen Yogurt. *Foods*, *13*(5), 808. DOI: 10.3390/foods13050808 PMID: 38472921

Chi, F., Tan, Z., Wang, Q., Yang, L., & Gu, X. (2024). Tibetan Butter and Indian Ghee: A Review on Their Production and Adulteration. *Agriculture*, *14*(9), 1533. DOI: 10.3390/agriculture14091533

Dineva, K., & Atanasova, T. (2023). Health Status Classification for Cows Using Machine Learning and Data Management on AWS Cloud. *Animals (Basel)*, *13*(20), 3254. DOI: 10.3390/ani13203254 PMID: 37893978

Fernández González, A., Badía Laíño, R., Costa-Fernández, J. M., & Soldado, A. (2024). Progress and Challenge of Sensors for Dairy Food Safety Monitoring. *Sensors (Basel)*, *24*(5), 1383. DOI: 10.3390/s24051383 PMID: 38474919

Gehlot, A., Malik, P. K., Singh, R., Akram, S. V., & Alsuwian, T. (2022). Dairy 4.0: Intelligent Communication Ecosystem for the Cattle Animal Welfare with Blockchain and IoT Enabled Technologies. *Applied Sciences (Basel, Switzerland)*, *12*(14), 7316. DOI: 10.3390/app12147316

Germani, L., Mecarelli, V., Baruffa, G., Rugini, L., & Frescura, F. (2019). An IoT architecture for continuous livestock monitoring using LoRa LPWAN. *Electronics (Basel)*, *8*(12), 1435. DOI: 10.3390/electronics8121435

Global Dairy Industry and Trends - Feed & Additive Magazine. (n.d.). Retrieved November 26, 2024, from https://www.feedandadditive.com/global-dairy-industry-and-trends/

Gunturi, M., Kotha, H. D., & Srinivasa Reddy, M. (2018). An overview of internet of things. *Journal of Advanced Research in Dynamical and Control Systems*, *10*(9), 659–665.

Haug, A., Høstmark, A. T., & Harstad, O. M. (2007). Bovine milk in human nutrition–a review. *Lipids in Health and Disease*, *6*(1), 1–16. DOI: 10.1186/1476-511X-6-25 PMID: 17894873

Hu, T., Zhang, J., Zhang, X., Chen, Y., Zhang, R., & Guo, K. (2023). The development of smart dairy farm system and its application in nutritional grouping and mastitis prediction. *Animals (Basel)*, *13*(5), 804. DOI: 10.3390/ani13050804 PMID: 36899660

Ionescu, A.-D., Cîrîc, A. I., & Begea, M. (2023). A review of milk frauds and adulterations from a technological perspective. *Applied Sciences (Basel, Switzerland)*, *13*(17), 9821. DOI: 10.3390/app13179821

Kalamaras, S. D., Tsitsimpikou, M.-A., Tzenos, C. A., Lithourgidis, A. A., Pitsikoglou, D. S., & Kotsopoulos, T. A. (2024). A Low-Cost IoT System Based on the ESP32 Microcontroller for Efficient Monitoring of a Pilot Anaerobic Biogas Reactor. *Applied Sciences (Basel, Switzerland)*, *15*(1), 34. DOI: 10.3390/app15010034

Mu, F., Gu, Y., Zhang, J., & Zhang, L. (2020). Milk source identification and milk quality estimation using an electronic nose and machine learning techniques. *Sensors (Basel)*, *20*(15), 4238. DOI: 10.3390/s20154238 PMID: 32751425

Neethirajan, S. (2023). Artificial Intelligence and Sensor Technologies in Dairy Livestock Export: Charting a Digital Transformation. In *Sensors* (Vol. 23, Issue 16). DOI: 10.3390/s23167045

Pandey, P., Pandey, A., Yan, L., Wang, D., Pandey, V., Meikap, B. C., Huo, J., Zhang, R., & Pandey, P. K. (2021). Dairy waste and potential of small-scale biogas digester for rural energy in India. *Applied Sciences (Basel, Switzerland)*, *11*(22), 10671. DOI: 10.3390/app112210671

Pu, Y., Pérez-Marín, D., O'Shea, N., & Garrido-Varo, A. (2021). Recent advances in portable and handheld NIR spectrometers and applications in milk, cheese and dairy powders. *Foods*, *10*(10), 2377. DOI: 10.3390/foods10102377 PMID: 34681426

Salah, B., Alsamhan, A. M., Khan, S., & Ruzayqat, M. (2021). Designing and Developing a Smart Yogurt Filling Machine in the Industry 4.0 Era. *Machines (Basel)*, *9*(11), 300. DOI: 10.3390/machines9110300

Scudino, H., Guimarães, J. T., Ramos, G. L. P. A., Santillo, A., Pimentel, T. C., Cruz, A. G., & Albenzio, M. (2023). High-intensity ultrasound in cheese processing. *Current Opinion in Food Science*, *50*, 101004. DOI: 10.1016/j.cofs.2023.101004 PMID: 37087267

Shao, Y., Yuan, Y., Xi, Y., Zhao, T., & Ai, N. (2023). Effects of Homogenization on Organoleptic Quality and Stability of Pasteurized Milk Samples. *Agriculture*, *13*(1), 205. DOI: 10.3390/agriculture13010205

Smith, N. W., Fletcher, A. J., Hill, J. P., & McNabb, W. C. (2022). Modeling the contribution of milk to global nutrition. *Frontiers in Nutrition*, *8*, 716100. DOI: 10.3389/fnut.2021.716100 PMID: 35096919

Su, J.-J., Ding, S.-T., & Chung, H.-C. (2020). Establishing a smart farm-scale piggery wastewater treatment system with the internet of things (IoT) applications. *Water (Basel)*, *12*(6), 1654. DOI: 10.3390/w12061654

Tangorra, F. M., Buoio, E., Calcante, A., Bassi, A., & Costa, A. (2024). Internet of Things (IoT): Sensors Application in Dairy Cattle Farming. *Animals (Basel)*, *14*(21), 3071. DOI: 10.3390/ani14213071 PMID: 39518794

Zadeike, D., & Degutyte, R. (2023). Recent advances in acoustic technology in food processing. *Foods*, *12*(18), 3365. DOI: 10.3390/foods12183365 PMID: 37761074

Żbik, K., Onopiuk, A., Górska-Horczyczak, E., & Wierzbicka, A. (2024). Trends and Opportunities in the Dairy Industry: A2 Milk and Processing Methods. *Applied Sciences (Basel, Switzerland)*, *14*(15), 6513. DOI: 10.3390/app14156513

Chapter 5
Integration of IoT in Dairy Facilities

Ahmed M. Alhussaini Hamad
https://orcid.org/0000-0001-5037-9379
Benha University, Egypt

Dina A. B. Awad
Benha University, Egypt

ABSTRACT

The integration of internet of things (IoT) technologies into dairy processing facilities has the potential to revolutionize the industry by improving operational efficiency, product quality, and safety. This chapter explores the current state of IoT adoption in the dairy sector, focusing on the key technologies, benefits, challenges, and future trends. IoT sensors and actuators enable real-time monitoring and control of critical process parameters, whereas data analytics and machine learning algorithms facilitate predictive maintenance, quality control, and process optimization. Future trends and innovations in this field include the development of advanced sensors, improved data handling capabilities, and the integration of artificial intelligence for predictive analytics and decision support. To fully harness the potential of IoT in the dairy industry, strategic planning, collaboration among stakeholders, and continuous investment in research and development are essential.

INTRODUCTION

The dairy industry today is continuously required to meet tough demands. Modern technological advancements are releasing new solutions in top gear to upscale (Fidan et al., 2023). In this information age, the term e-dairy illuminates the evolution of

DOI: 10.4018/979-8-3693-9949-1.ch005

new technology. The primary aim of this era is to maintain the Internet of Things (IOT) with respect to the dairy farm. IoT is estimated to cause propagation in a very fast manner in real-time, and in a few decades, would lead to a revolution of progress in the economy (Kavitha & Ravikumar, 2021).

Incorporation of measures of the Internet of Things in the dairy processing plants has become an important aspiration for the dairy industries as technical support, as the operational pattern of the dairy industry has been irregular and dynamic (Jerhamre et al., 2022; Gowda et al., 2024). Most of the current processes are in need of these kinds of technologies. These technologies make it operational to adopt a new pioneering approach for the control of operations. Incorporation of the use of IoT is not as smooth as it is being visualized since the chances of errors are high. The effective and efficient function of the IoT operations is necessary for the efficient and effective data extraction systems (Jadhav et al., 2021). (Protopsaltis et al.2020) (Jadhav et al.2020). The current technologies of the IoT with respect to the entirety are needed to be incorporated, and discussion is necessary to deal with managing repair complexity systems. Product-centric change management operations and product control operations are needed for the effective operation of milk (Varma et al., 2021). Keeping in view the preceding issues, it is necessary to delve into information to an optimal extent to realize different interrelated issues of mechanisms of data extraction, crucial technological mechanisms of IoT, and its associated systems operations with interrelated aspects.

Figure 1. IoT integration in dairy processing sector

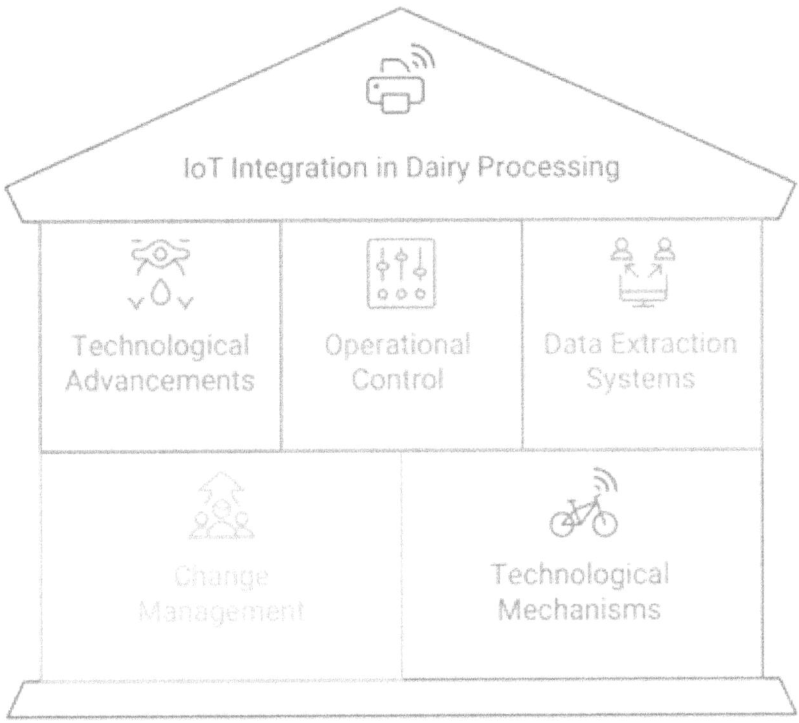

A variety of methods of milk production and processing were developed in the pre- and post-Industrial Revolution eras. From traditional instruments and rudimentary, time-consuming procedures, the sector had advanced significantly by the early twentieth century because of breakthroughs in steam and modern electricity (Groumpos, 2021; Sahoo & Lo, 2022). The industry has also seen enormous growth in automation. Technology has played a significant role in providing higher quality, more quickly, and less expensive goods. Dairy farming has commonly been a slow-to-change sector, even though the industry already utilized current technology to streamline some processes. ((Dimitri & Nehring, 2022; Rathakrishnan et al., 2022). Charging the cost, some organizations shied away from spending on anything different and left business for the following month. Despite many dairy owners wanting to develop their processing ability or undertake brands as a way to raise their domestic income (Bunce, 2020; Al Sidawi et al., 2021).

An IoT-based network of equipment applications can conduct and handle the procedures. It can compute activities like preparing schedules and predicting the completion period of procedures; secure the plant surroundings by managing farm hands' eight-hour shifts; identify illegal energy surges used by the plant's power network; and gain deep insight into extensive production of some food products utilizing preselected and disciplined input variables for the facilities. Adding the sensor together would enable us to monitor major pieces of equipment such as vats, dryers from anywhere in the area, or lactoplast aromatic panels and pallet temperature of the case. Increased levels of research and significant operating difficulties for agriculture have increased the urgency for this transition (Avhale et al., 2021; Syrmos et al., 2023). For example, few regulations have a large effect on their value-added milk processing. These regulations are everywhere, i.e., at least in the U.S. and EU level, and postpone Japan's income as importers. We aim to raise our test to a large and growing industry. Concerns around quality and protection have pursued it, and people regard dairy as a means to link a story to distant lands. Aside from people, numerous pet dairy brands to the north of the U.S. advertise as 'New Zealand Design.' They satisfy buyer desires for trust in quality and freshness (Syrmos et al., 2023).

IOT TECHNOLOGIES IN DAIRY PROCESSING

Internet of Things (IoT) technologies are products of the digital revolution that are already reshaping the dairy processing and production industries. These technologies can be found across the entire dairy processing supply chain. At the level of dairy processing facilities, two critical components that are of interest are sensors used to measure some variable over time and actuators used to change some state of the system being monitored and controlled (Akhigbe et al., 2021; Neethirajan, 2023). The data collected by these devices can be harnessed in real time to monitor and control a wide variety of processing steps from feeding and milking to chilling, pasteurization, and storage. This data can also be used to achieve increased precision in forecasting payout levels and to thereby correctly forward-contract inputs and live cattle (Alonso et al., 2020; Burmistrov et al., 2021). This intelligence can have a profound effect on the way resources are managed, and waste is minimized. Moreover, data analytics and machine learning powered by these devices can greatly enhance the power of predictive laws in operations management to predict and manage customer demand and initiate just-in-time operations (Helo & Hao, 2021; Ivanov et al., 2021; Pasupuleti et al., 2024). This has the potential to revolutionize the business logic of existing dairy processes.

At its simplest, a dairy processing plant that has fully embraced IoT will use many interconnected devices, such as mobile phones, sensors, and actuators in centered operations to make operational decisions in real time. This interconnected network of devices will feed its big data architecture with minute-by-minute real-time operational, tactical, and strategic information (Michie et al., 2020; Cabrera

& Fadul-Pacheco, 2021). The facility can utilize specialized machine learning algorithms to run these big data analyses in order to make decisions. To truly encapsulate a connected processing facility, all processing equipment must be built with interoperability in mind. This should encompass every nut, bolt, screw, and head in production from raw material intake machines through chillers to filling machines, inventory hold rooms, and refrigerated transport systems (Bryndin, 2020; Ordieres-Meré et al., 2023).

Figure 2. Application of IoT in dairy sector

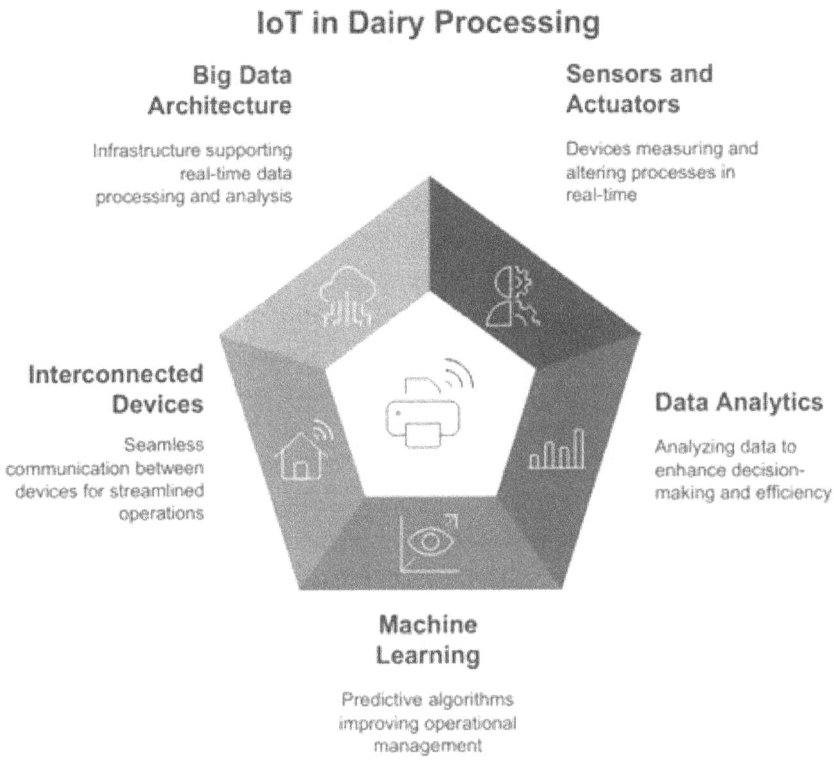

Sensors and Actuators

The first element of modern dairy processing operations and essentially any processing industry with dairy processing can be called 'transparency of the process'. This means that each step of a dairy processing operation can be visualized and inspected not only by an operator, but also by a quality manager, a maintenance manager, an auditor, a manager of the company, and even a dairy consumer (Pu et al., 2020; Malik et al., 2024). To achieve this transparency of operations, a sufficient number and type of sensors and actuators are needed, which are connected to systems. It should be stated that transparent processing operations do not mean a ready-to-use strategy to maximize performance and minimize energy and raw materials use. However, full transparency of process operations is a prerequisite for managing the processing system (Venkatesh et al., 2020; Lohmer & Lasch, 2020). With good system management, more efficient and effective processing of produced foods can also be realized. Both transparency and effective management imply that data integrity will be high (Rejeb et al., 2020; Leal et al., 2021; Feng et al., 2020).

Sensors integrated with the system provide real-time data of certain parameters, some of which are directly linked with the safety of the product and can be called a critical control point. The sensed and streamed dataset will allow the assessor to recognize if the process parameters are moving in the right direction or if manual intervention is needed at that point (Pu et al., 2020; Ziyaina et al., 2020; Kyaw et al., 2024). Simultaneously, the process will become more predictable and manageable, as with knowledge of the past, far more tailored automatic actuators can be built. As a result, this will lead to a substantial improvement in efficiency along with higher confidence in how the process operates. Measurement of quality parameters is an essential data source for the operator. However, data must be accurate, reliable, and available in real-time in order to make the right corrective decisions (Novaes et al., 2020; Naqvi et al., 2021). In addition, it is crucial for consumers to know that appropriate authorities have access to this data to safeguard the products that are processed. Some quality measurement operations are based on combinations. For instance, the assessment of organic milk quality can partly be app-based, enabling farmers to focus on quality. In the milk processing chain, several sensors are used to monitor the presence of impurities such as antibiotics in milk (Nagraik et al., 2021; Lal et al., 2022; Sowmya & Ponnusamy, 2021).

Figure 3. Enhancement of food safety and quality using IoT

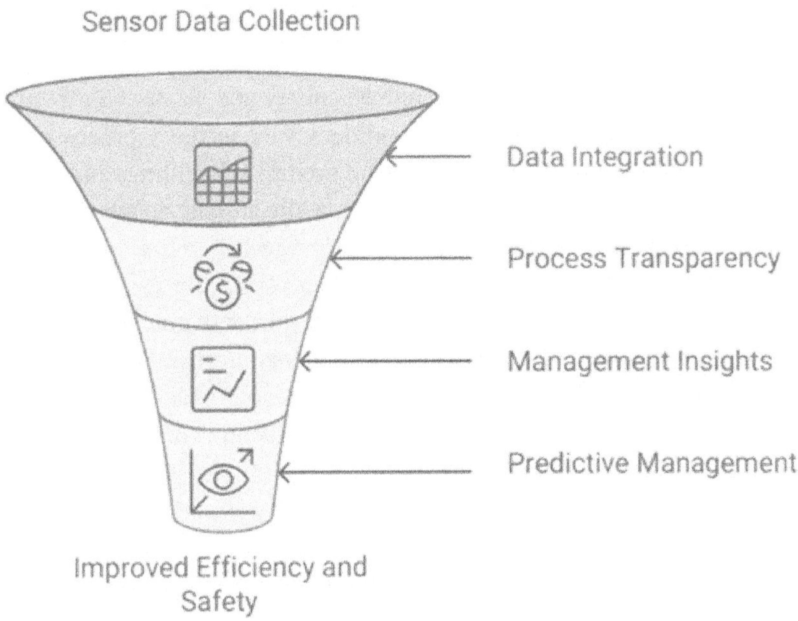

Data Analytics and Machine Learning

As stated in Section 2.1, one basic function of IoT in the dairy processing industry is the collection of data. Data produced by IoT devices are raw, unrefined, and of little use to decision-making, but by applying data analytics, this raw data can be transformed into usable and valuable information such as insights and/or decisions. Data analytics allows dairy facilities to analyze historical data, mainly reflective of actions (such as adjustments in case of significant deviations), and perhaps spot trends (Gehlot et al., 2022). A method that goes a step further in this process is the application of machine learning. One of the use cases was described in Section 2.3, where it is applied in predictive maintenance. Equipment behavior and measurements are used to predict failure before it actually happens, preventing or reducing downtime. IoT, together with data analytics and machine learning, can also be used to better understand the causes of failures and deviations using advanced algorithms and discover new sensors and attributes that correlate highly (Cockburn, 2020; Wu & Hsiao, 2021; Chi et al., 2022).

Further expansion is the application of machine learning algorithms to predict when final quality measurements can be taken. In this manner, deadline-based production processes can be fine-tuned to reduce energy and wastewater treatment costs. This would allow dairy industry stakeholders, such as dairy processors and farmers, to predict the product's final quality (and process scheduling) and duration of shelf life (and subsequent distribution network), while gaining an advantage over the competition based on product innovation. The project is in the middle of surveying the European dairy sector to identify whether such a use case would be of interest and can fill the needs of the industry. Besides product quality and process innovation, data analytics can be used to ensure quality control and uniformity, and possibly food safety by implementing advanced process control and optimization of critical control points (Konstantinidis et al., 2023)(Hassoun et al., 2024) (Chowdhury et al., 2025). Additionally, it can be part of traceability systems in evaluating how and where a product failed, and whether this failure was part of a larger system malfunction or subpar performance. To implement data analytics capabilities, the industry will face the same challenges as IoT data collection. A proper system that can continuously monitor the data, alarm when significant deviations occur, manage the data, and securely store it is required. With the steady increase in the volume of data produced, the challenge of interpreting sensor signals augmented into complex datasets is needed (Choi et al., 2021; Theissler et al., 2021; Kreuzberger et al., 2023). The staff monitoring the performance of the sensors and reacting to physical phenomena will require proper training to understand the signals. Environmental data retrieval and knowledge of data analysis software for personalized urban and industrial applications based on sensor technologies will become important. Domain expert developers who can enhance the usability and usefulness of the platforms and tools do not need to be trained in the development of sensors and detection algorithms but only trained in software development. Data analytics edge: operational efficiency, innovation, and developing data-driven collaborative business models. A data-driven decision model must also enable, stimulate, and accompany the sharing of valuable insights on emerging questions, desired innovation on food products, and technological developments in dairy processing. The industry is experiencing scanners and other collection methods needed to observe internal processes. Models are also needed and built to assess the quality and shelf-life assessment (Newton et al., 2020; Fisher et al., 2020; Baldin et al., 2021; Cabrera & Fadul-Pacheco, 2021).

Figure 4. IoT in dairy sector components

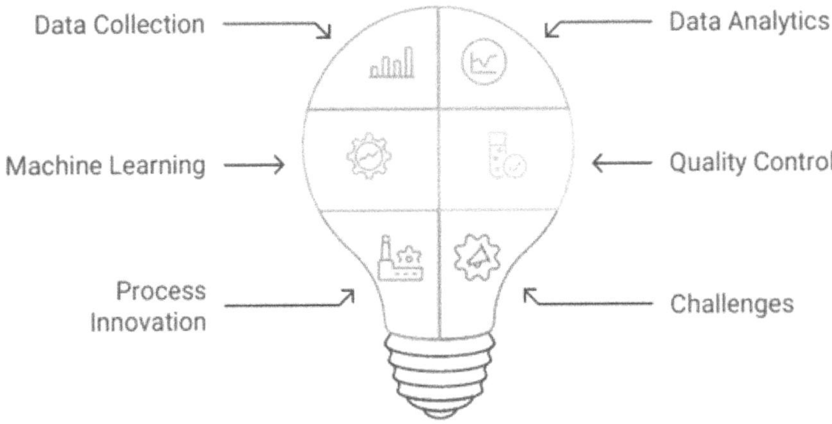

BENEFITS OF IOT INTEGRATION IN DAIRY PROCESSING

Benefits of IoT in Dairy Processing

In general, the most commonly cited benefits of IoT in a dairy processing facility are grouped into the following five categories: quality, waste and resource management, compliance, and traceability. Firstly, there are clear quality benefits to being able to track batch or lot-specific data at each processing step. In terms of consistent raw milk parameters, monitoring the data can improve quality control. Advances in animal feeding are likely to occur, and IoT technologies will make consistent data collection across herds inside a farm an easy task. Milk, and the cows that produced it, can be tracked inside the processing facility, and a premium price can be given to the milk produced at a farm with the desired data (Tan & Ngan, 2020; Khanna et al., 2022).

The grading and quality of the milk can be constantly monitored by plant staff with real-time data coming through the dairy processing chain. Integration of all data sources and trackability with the recording of processing data at critical control points can contribute to improving cheese production practices. In addition, the company can peg most of their production and staffing levels around current averages or previous selling records. The detailed data can also be linked with herd management principles to reduce waste and streamline operations inside the plant. In general, less waste in consumer products means either less reprocessing costs or added value in products shifted back into the higher-paying derivative by further processing at a slower line speed (Murphy et al., 2021). (Nyokabi et al., 2021;Neculai-Valeanu & Ariton, 2022).

There are also benefits in stock and resource management, particularly in water management. Reduced detection time for water and effluent leaking inside a plant based on sensor locations will guide the cleanout of containment systems, reducing cross-contamination risk or fines. This extends further to the size of 'live' biological data. Industries are always being exposed to audits and frequent ongoing assessments. Whichever company can collect direct data from the product at the site of production and afford the customer a scanner link to view it; the buyer can see this to make sure that it meets the requirements they have stipulated. Reducing in-person presence makes compliance easier. Improved process and product tracking will be benchmarked against central agencies' interventions and thus deliver improved global trade relationships. In the dairy processing sector, a single cow's milk quality can differ moment to moment, and IoT will enable the use of ruminant farm profile information to assist in modifying the operations to the task at hand (Michie et al., 2020; Alonso et al., 2020; Gehlot et al., 2022).

Figure 5. Integration of IoT in dairy sector

Improved Quality Control

The introduction of IoT technologies can have multiple impacts on the operation of dairy processing systems. This subsection will describe the role of IoT sensors in quality control of dairy processing facilities. Constant monitoring of environmental conditions allows dairy product manufacturers to ensure that their product is well within the safety standards defined by regulatory bodies at all times. Monitoring the product during processing is also essential in dairy applications (Owusu-Kwarteng et al., 2020; Ledo et al., 2020; Gehlot et al., 2022; Neethirajan, 2023). Contamination of the product must be identified as quickly as possible to ensure timely corrective action. For this reason, real-time monitoring is preferred. Real-time monitoring of milk or milk-processed products to ensure the process parameters such as acidity, dissolved oxygen, temperature, etc., stay within a certain range is essential to keep the possibility of contamination at all times low, regardless of the stage of processing. IoT sensors are essential in this type of real-time monitoring due to their high measurement accuracy, high data transmission rate, and a large range of parameters that they can measure simultaneously (Alonso et al., 2020;Lal et al., 2022; Hadi et al., 2024)

IoT sensors have an embedded system for the storage of verified data. This pack of verified data creates a continuous historical log ensuring reliable tracking as and when required. During any form of real-time monitoring and regulation of the product, movement history can be combined, thus enabling a timely decision process. Due to improved control in the supply chain, governments are pressing more on quality and traceability due to food safety regulations, which are in line with the philosophy (Iftekhar & Cui, 2021). An unexpected quality error can, more often than not, lead to a product recall. The faster the root cause is identified, the less the negative impact due to product recall. Brand owners can protect bottom-line estimates in the wider chain. Consumers look for dairy products that are safe and consistent. With their premium purchase, they expect the same taste profile from dairy products (Soon et al., 2020; Soon & Abdul Wahab, 2021). Quality that used to be considered a brand differentiator by the producer is now becoming a basic expectation of consumers, resulting in greater competition in the world of dairy processors. Without a commitment to quality, it becomes difficult to compete. In essence, IoT sensors are utilized mainly in quality control in dairy processing facilities and in retrofitted storage and transportation of milk (Burmistrov et al., 2021).

Figure 6. Integration of IoT in quality control in dairy processing

Enhanced Production Efficiency

In the context of dairy, the advancement of the Internet of Things (IoT) can enhance the efficiency of production to a large extent. The production process in these plant operations faces various unexpected delays, which could hamper the timely production of end products. The integration of real-time monitoring of machines, support utilities, and raw materials will make data accessible, thereby providing better decisions for production managers (Alonso et al., 2020; Pu et al., 2020; Cabrera & Fadul-Pacheco, 2021). The ease of data logging, switching between product lines, and automated inflow of raw materials will enable the production facility to optimize production scheduling, ordering of raw materials, and maintenance scheduling on a real-time basis. Some of the IoT applications will drastically influence an enhancement in the efficiency of dairy processing plants (Akhigbe et al., 2021; Nozari et al., 2021;Gehlot et al., 2022). Real-time monitoring and control in dairy processing plants can be used to take corrective actions in case of any failure, thereby avoiding breakdowns and helping to increase the availability of sensitive machines used for specific value-added operations. Also, as per the measurement and verification of various set points, the energy savings are quantified and indicated (Wangen et al., 2021; Heema et al., 2022; Malik et al., 2024). The IoT hardware learns the power rating of the machine over a period of time and signals an alarm to the operator in case the power consumption exceeds the rated amount. The instantaneous values of voltage, current, and power for each machine are monitored consistently through IoT, which will help to assess harmonic studies. Machines like feed mills, boilers, etc., give indications of the wear and tear of their parts through noise variations on bearings, elongated start-up times, and non-optimized power consumption. These vibration shakers provide an early alert to the maintenance team. Predictive maintenance reduces delays due to breakdowns, which the maintenance team fixes at their convenience. The Overall Equipment Effectiveness (OEE) of machinery is calculated in real time and displayed on the screen for production managers to optimize utilization based on the production plan (Novaes et al., 2020; Pancholi et al., 2023; Uzoigwe, 2024). (Shahin et al., 2023).

Automation of certain processes in processing, packaging, and loading and dispatch will avoid numerous delays. For example, front loading of raw milk tankers can be done with the help of terminals where the material handler can receive a signal from the stop button in advance. The IoT hardware will direct incoming trucks to the proper lanes available for operations. Automating the following activities will drastically reduce the time for modification and avoid delays: front loading, steaming, homogenizing, pasteurizing, and refrigeration of raw milk; automated changeovers across different stock keeping units of finished products; automated addition of support utilities such as steam and water with the products; directives to trucks for back loading, including sequencing of the products (Ghosal et al., 2020; Moszyk et al., 2021; Abourraja et al., 2022; Najafi et al., 2024).

The above are just a few listed items in IoT hardware, where one can achieve direct savings not only in efficient decision-making capabilities but also in terms of margins from the machinery and improvement of productivity versus distress time and associated manual fixes. Additionally, apart from direct savings, the implementation of advanced technology is moving from operating systems to Enterprise Risk Management in today's era. Hence, due to the possibility of easy integration with IT equipment, it has the highest growth potential compared to other available technologies (Nozari et al., 2021;Carrillo-Lopez et al., 2021; Sar et al., 2022; Mehany et al., 2024).

Figure 7. How can IoT help in dairy sector enhancement?

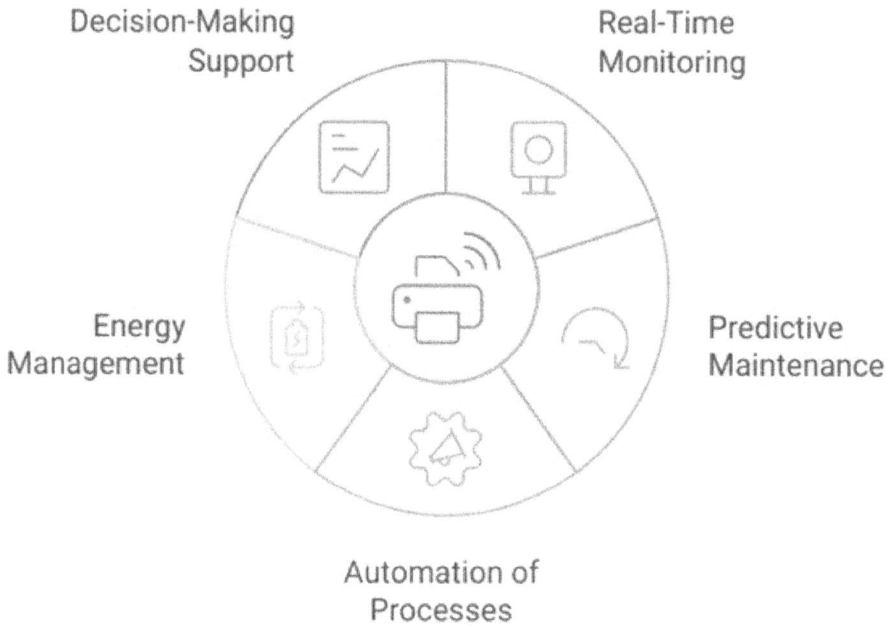

CHALLENGES AND LIMITATIONS

The dairy processing infrastructure and the methods have been undergoing radical changes. Some process controls have been implemented in some dairy plants; typically, they are local control systems (Beber et al., 2021). The efficiency of the control of the process plant can be greatly increased by the interconnection of different devices. Low-cost sensors are being developed using technology, which can be helpful in determining the critical quality attributes of food (Vejan et al., 2021; (Li et al., 2021). Although this technology is already successfully applied in some sectors, the process of finding its way into the general dairy industry is still in its initial stages. A number of challenges and limitations associated with integrating these systems in smart dairy processing facilities are discussed (Alonso et al., 2020).

The challenges in practical implementation include the complexity of implementing the systems, based on the interoperability of a range of multiple sensor devices and compatible networking protocols (Hazra et al., 2021; Lee et al., 2021). Comprehensive data security measures need to be put in place for authentication and confidentiality of the data. It is widely expected that the costs of using these advanced technologies need to be carefully assessed for comprehensive cost-benefit analysis, particularly for smaller dairy producers. This, therefore, needs strategic planning by the dairy companies, investment across the value chain, and support from the regulators. It will also require dairy industry workers to develop skills in managing and interpreting the data. There is also an issue of regulatory compliance for companies that want to use these new advanced technologies (Mrabet et al., 2020; Manivannan et al., 2020). Regulatory authorities normally take time to update guidelines while these advanced technologies are rapidly advancing. It is, therefore, necessary to develop and review the regulatory approach in view of the evolution of technology. Ethical criticism about how technology will affect society and humankind, the mass workforce of workers, etc., needs to be considered (Khanna et al., 2022)(Neethirajan, 2023). Established food safety standards and the increase of consumer confidence gain with these technologies, so that appropriate recognitions can be disseminated throughout dairy processors. The system must be user-friendly and cost-effective for realizing objectives, and hence it needs a comprehensive cost-benefit analysis. In this system, protection of sensitive data is essential. To meet the challenge, a proper and efficient cybersecurity system is needed. The rise and explosion of devices provide a field day for hackers (Kyaw et al., 2024). Due to technical advancement, the development of appropriate cybersecurity for the dairy industry becomes mandatory. In light of the growing need for advanced technology and sensor devices in the dairy industry, this serves as a guide for stakeholders who are considering this approach to dairy product production. Working towards this approach would mitigate the main challenges and limitations posed by the adoption of this technology (Nikander et al., 2020). (Kjønås & Wangen, 2024).

Data Security Risks

Data security risks are increased when IoT technologies are utilized in facilities. One of the major concerns in this area is that every connected device becomes a potential entry point for attackers. Because IoT devices lack the necessary protection measures, they can be more easily compromised to cause harm (Iftekhar et al., 2020; Alonso et al., 2020; Demestichas et al., 2020. Akhigbe et al., 2021). Thus, an entire network may be easily disturbed. When information is transmitted from one device to another over the internet, any financial loss from the exposed information falls solely on the organization. Meanwhile, leaked information may be beneficial to other facilities, causing a significant blow to a facility's wide-reaching reputation. The execution of data safety protocols can significantly mitigate these risks associated with this market (Ledo et al., 2020). This encompasses the implementation of strong passwords, two-factor authentication, real-time controls, better database configuration, regular searches, operation restrictions, updates, full testing, encryption, and wiretap prevention. Similarly, every person must be aware of potential risks since intruders may try to access a network or a system (Wu & Hsiao, 2021; Khanna et al., 2022).

Employees and facilities must thus enhance their cybersecurity awareness in order to reduce the potential risks and outcomes. As a result, the worker community needs routine digital training. Consequently, every worker needs to be trained and instructed in the necessary actions they ought to take to maintain processing equipment and apply the aforementioned safety steps. This additional training allows companies to reinforce the protective practices that govern the operation of processing facilities. In addition, it may foster a more secure atmosphere. This methodology emphasizes the importance of imparting knowledge of digital technology's operation in the staff. In conclusion, while IoT is a paradigm of multiple aspects, the potential complications necessitate the implementation of proper safety procedures to ensure that our products are stored in an effective manner (Sinha & Dhanalakshmi, 2022).

Integration Complexity

Complexity in integrating IoT in a dairy processing facility can arise from the fact that the devices used in the processing operation belong to different manufacturers, which may use a closed platform for the deployed hardware and related software or for their data integration software solution that is already functioning in the facility. Each manufacturer needs to develop its own software applications so that they can work seamlessly with new devices across operating systems. Often, multiple interfaces to devices are required, with development work specific to each device and limited reusability of tools (Trakadas et al., 2020; Schiller et al., 2022; Theodorakopoulos et al., 2024). The result is that the time and cost of development work can rise significantly. Existing technologies can make it difficult to correlate data monitoring. Change is needed in existing data management and analytics, requiring new, extended, and/or upgraded information technology infrastructure. Knowledge and sensor readings are already hard to interpret, and limited access to specialist sources can make it more difficult for plant personnel to effectively perform their day-to-day work. Adoption of IoT devices will either add to the complexity of sensors or take some existing conditions (e.g., humidity, vibration, temperature, conductivity) and use algorithms to predict future behavior based on these conditions (Gupta & Quamara, 2020; Pappas et al., 2021;Allioui & Mourdi, 2023).

The use of IoT would allow plant personnel to track the entire data pattern required for the sensor to function, but only to report on the sensor conditions. This approach can be time-consuming for plant personnel to adopt and then express these reports in a meaningful way. The existing IoT device solution is seen as adding little value that does not moderate control (Taheri et al., 2021; Morais et al., 2021;

Abdel-Basset et al., 2024). Moreover, the plant personnel had to be trained to use and handle new, complex IT-based tools regularly, and a support contract with vendors should be established; availability of the IoT solution was generally for vital business instrumentation only. As a result, advance planning indicates that existing plants are likely to have limited quick wins for the integration of IoT devices, except for data connectivity between relatively simple/legacy controllers and existing data exchange servers (Moudgil et al., 2023). Likewise, system upgrades are identified that would be required to allow integrations with advanced warehouse management systems. Overall, the potential impact of any major delays in execution is likely to remain quite low while IoT integration solutions are not yet available and their long-term viability and stakeholders are unclear (Nižetić et al., 2020). A detailed IoT strategy for the complete production chain, once these integrated IoT systems are required, would be needed. Ensuring these sophisticated systems do not unduly disrupt present operations and that they will be capable of meeting the operational needs and expectations of the entire company plant from day one would also require a detailed stakeholder communications approach (Attaran, 2020; Ivanov et al., 2021; Belhadi et al., 2022).

Figure 8. Challenges facing IoT integration in dairy processing sector

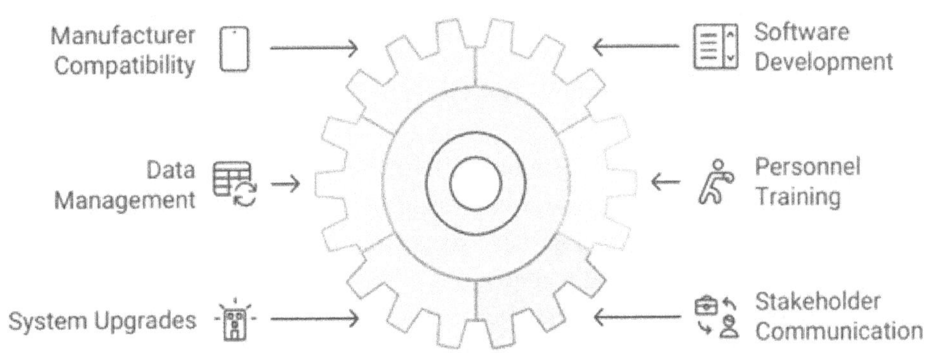

SUCCESSFUL IMPLEMENTATIONS OF IOT IN DAIRY PROCESSING

The previous sections showcased challenges and some IoT solutions that support and solve them. Below are two interesting IoT implementations suitable for processing sites, to give another set of insights into this area. These cases bring great advancements in operations and product quality to their users. Drupe is a client of Sensory Dimensions and was responsible for the human panels and consumer research discussed in this article. Dairy cooperative Arla decided to run a pilot at one of their dairies in the UK to investigate whether new technology could be used to indicate the 'creaminess' of cream style products online, providing information that could be correlated with a sensory attribute. After specifications were written with input from a sensory expert, a system was commissioned from Parkend as part of an upgrade to the cream processing plant.

A number of technologies to produce IoT devices in the processing plant already existed at the domestic level or outside of the dairy industry. But every sensor tailored to the cream process at Parkend needed to be fresh and bespoke. They had to be rugged enough to stand up to being installed on the cream and skim reservoirs, and the equipment would need to be able to be validated and user calibrated to use the sensor for cross-plant and inter-dairy work. In 2019, a test unit ready for trial was operational in the Parkend factory, providing information to the scarper operators in frequency and table action protection to ensure an even, repeatable, and good product to the consumer of high yield therefrom. 'Once this trial has been done, we will be gathering the process and product data that will enable us to better understand the opportunity we can deliver in choosing the right sensor for the high-end market.' (Hepfer & Lawrence, 2022).

FUTURE TRENDS AND INNOVATIONS

This ongoing era of digital technology, with the proven success of IoT across different domains, is heading towards gearing up its capabilities for delivering efficient and effective solutions in dairy processing sectors (Michie et al., 2020). The current methodologies and technologies of IoT will continuously advance toward proactive, automated, and intelligent services that customers can use in dairy processing facilities (Akhigbe et al., 2021). There are some scenarios and trends that are going to become innovations in the dairy processing sector when combined with IoT technologies, such as the development of sensors that provide actionable information, improving data handling capabilities like data mining, predictive analytics, and sustainable interoperability with AI. The role of AI combined with IoT technologies will be able to predict future probabilities based on data that have never happened, but still, their probability of happening is more than 75% based on the analytics of a series of historical data (Cabrera & Fadul-Pacheco, 2021; Neethirajan, 2023; Konstantinidis et al., 2023).

Figure 9. Future innovations in dairy sector

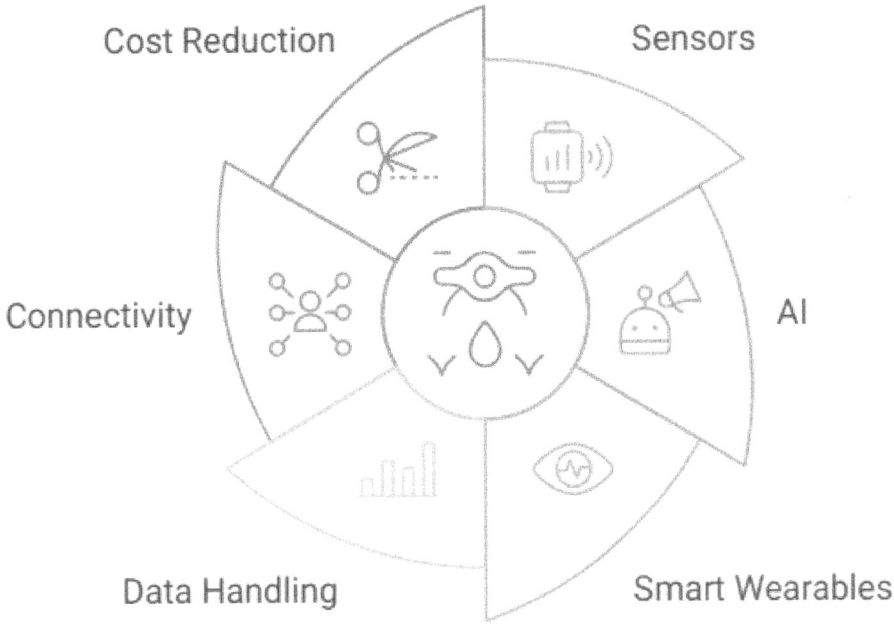

Based on the outcome of various trends and issues and the influence of IoT developments in the related field, the technological innovations that are going to happen in the field of dairy processing facilities in the coming years include smart wearables for cows, endurance of IoT sensors, AI that enhances monitoring and user insight, milk volume sensors, inactive LED mass, and agronomic AI, perceptions-informed improved connectivity/interoperability, alerts, and communication. It is important that continuous investment in research and development be done (Neethirajan, 2023; Džermeikaitė et al., 2023). As a result, it is possible that properly integrated technologies not only reduce the operation and maintenance costs of systems but also can increase the reputation of the processing facility from customer insights, anticipating what they actually need by showing usable qualitative notifications and details about the product in the near future with better-predicted information during the alliance (Gehlot et al., 2022; Džermeikaitė et al., 2023; Alipio & Villena, 2023).

CONCLUSION AND RECOMMENDATIONS

The benefits of IoT technologies in dairy processing facilities are perceived to be transformational and are expected to result in improved operational efficiencies and improved product quality and safety. The main challenges in implementing IoT solutions are associated with security and integration complexities. To develop, utilize, and innovate IoT within the dairy sector, strategic planning and long-term vision should be developed in collaboration with stakeholders. Both large milk processors and medium-sized dairy processing facilities see value in implementing IoT technologies in their operational processes (Alonso et al., 2020). The main drivers of implementing IoT within dairy processing facilities are related to the potential for overseeing replaceable parts and production equipment to limit unscheduled downtime (Marchi et al., 2024). The dairy sector is facing significant changes; opportunities for growth and job creation will be linked to the capacity of industry stakeholders to embrace new technologies such as IoT. Investment in stakeholder training and the retention of talent is one of the key criteria for success in the industry. IoT technologies and systems are becoming more sophisticated and at a faster rate than any other period in industry history. As such, it is important for the industry to proactively invest in a manner that allows it to learn what the most suitable applications are within our processing facilities that are reasonably future-proof to ensure competitiveness in the market of the future.

REFERENCES

Abdel-Basset, M., Hawash, H., Moustafa, N., Razzak, I., & Abd Elfattah, M. (2024). Privacy-preserved learning from non-i.i.d data in fog-assisted IoT: A federated learning approach. *Digital Communications and Networks, 10*(2), 404–415. DOI: 10.1016/j.dcan.2022.12.013

Abourraja, M. N., Kringos, N., & Meijer, S. (2022). Exploiting simulation model potential in investigating handling capacity of Ro-Ro terminals: The case study of Norvik seaport. *Simulation Modelling Practice and Theory, 117*, 102513. DOI: 10.1016/j.simpat.2022.102513

Akhigbe, B., Munir, K., Akinadé, O., Akanbi, L., & Oyedele, L. (2021). IoT Technologies for Livestock Management: A Review of Present Status, Opportunities, and Future Trends. *Big Data and Cognitive Computing, 5*(1), 10. DOI: 10.3390/bdcc5010010

Al Sidawi, R., Urushadze, T., & Ploeger, A. (2021). Factors and Components Affecting Dairy Smallholder Farmers and the Local Value Chain— Kvemo Kartli as an Example. In *Sustainability* (Vol. 13, Issue 10). DOI: 10.3390/su13105749

Alipio, M., & Villena, M. L. (2023). Intelligent wearable devices and biosensors for monitoring cattle health conditions: A review and classification. *Smart Health (Amsterdam, Netherlands), 27*, 100369. DOI: 10.1016/j.smhl.2022.100369

Allioui, H., & Mourdi, Y. (2023). Exploring the Full Potentials of IoT for Better Financial Growth and Stability: A Comprehensive Survey. In *Sensors* (Vol. 23, Issue 19). DOI: 10.3390/s23198015

Alonso, R. S., Sittón-Candanedo, I., García, Ó., Prieto, J., & Rodríguez-González, S. (2020). An intelligent Edge-IoT platform for monitoring livestock and crops in a dairy farming scenario. *Ad Hoc Networks, 98*, 102047. DOI: 10.1016/j.adhoc.2019.102047

Attaran, M. (2020). Digital technology enablers and their implications for supply chain management. *Supply Chain Forum: An International Journal, 21*(3), 158–172. DOI: 10.1080/16258312.2020.1751568

Avhale, G. V., Shaligram, A. D., & Gharpure, D. C. (2021). IoT based intelligent tracking, early-warning & management (ITEM) tool for efficient milk transportation in dairy industry. *AIP Conference Proceedings, 2335*(1), 60003. DOI: 10.1063/5.0043687

Baldin, M., Breunig, T., Cue, R., De Vries, A., Doornink, M., Drevenak, J., Fourdraine, R., George, R., Goodling, R., Greenfield, R., Jorgensen, M. W., Lenkaitis, A., Reinemann, D., Saha, A., Sankaraiah, C., Shahinfar, S., Siberski, C., Wade, K. M., Zhang, F., & Cabrera, V. E. (2021). Integrated Decision Support Systems (IDSS) for Dairy Farming: A Discussion on How to Improve Their Sustained Adoption. *Animals (Basel), 11*(7), 2025. Advance online publication. DOI: 10.3390/ani11072025 PMID: 34359153

Beber, C. L., Lakner, S., & Skevas, I. (2021). Organizational forms and technical efficiency of the dairy processing industry in Southern Brazil. *Agricultural and Food Economics*, *9*(1), 23. DOI: 10.1186/s40100-021-00195-3

Belhadi, A., Kamble, S., Fosso Wamba, S., & Queiroz, M. M. (2022). Building supply-chain resilience: An artificial intelligence-based technique and decision-making framework. *International Journal of Production Research*, *60*(14), 4487–4507. DOI: 10.1080/00207543.2021.1950935

Bryndin, E. (2020). Formation and Management of Industry 5.0 by Systems with Artificial Intelligence and Technological Singularity. *American Journal of Mechanical and Industrial Engineering*, *5*(2), 24–30. DOI: 10.11648/j.ajmie.20200502.12

Bunce, B. (2020). Dairy Joint Ventures in South Africa's Land and Agrarian Reform Programme: Who Benefits? In *Land* (Vol. 9, Issue 9). DOI: 10.3390/land9090328

Burmistrov, D. E., Pavkin, D. Y., Khakimov, A. R., Ignatenko, D. N., Nikitin, E. A., Lednev, V. N., Lobachevsky, Y. P., Gudkov, S. V., & Zvyagin, A. V. (2021). Application of Optical Quality Control Technologies in the Dairy Industry: An Overview. In *Photonics* (Vol. 8, Issue 12). DOI: 10.3390/photonics8120551

Cabrera, V. E., & Fadul-Pacheco, L. (2021). Future of dairy farming from the Dairy Brain perspective: Data integration, analytics, and applications. *International Dairy Journal*, *121*, 105069. DOI: 10.1016/j.idairyj.2021.105069

Carrillo-Lopez, L. M., Garcia-Galicia, I. A., Tirado-Gallegos, J. M., Sanchez-Vega, R., Huerta-Jimenez, M., Ashokkumar, M., & Alarcon-Rojo, A. D. (2021). Recent advances in the application of ultrasound in dairy products: Effect on functional, physical, chemical, microbiological and sensory properties. *Ultrasonics Sonochemistry*, *73*, 105467. DOI: 10.1016/j.ultsonch.2021.105467 PMID: 33508590

Chi, Y., Dong, Y., Wang, Z. J., Yu, F. R., & Leung, V. C. M. (2022). Knowledge-Based Fault Diagnosis in Industrial Internet of Things: A Survey. *IEEE Internet of Things Journal*, *9*(15), 12886–12900. DOI: 10.1109/JIOT.2022.3163606

Choi, K., Yi, J., Park, C., & Yoon, S. (2021). Deep Learning for Anomaly Detection in Time-Series Data: Review, Analysis, and Guidelines. *IEEE Access : Practical Innovations, Open Solutions*, *9*, 120043–120065. DOI: 10.1109/ACCESS.2021.3107975

Chowdhury, M. A. H., Reem, C. S. A., Rahman, S. M., & Sarkar, F. (2025). Comprehensive approaches for ensuring microbial safety in the dairy industry: Monitoring systems, inhibitory strategies and future prospects. *Food Control*, *168*, 110894. DOI: 10.1016/j.foodcont.2024.110894

Cockburn, M. (2020). Review: Application and Prospective Discussion of Machine Learning for the Management of Dairy Farms. In *Animals* (Vol. 10, Issue 9). DOI: 10.3390/ani10091690

Demestichas, K., Peppes, N., & Alexakis, T. (2020). Survey on Security Threats in Agricultural IoT and Smart Farming. *Sensors (Basel)*, *20*(22), 6458. Advance online publication. DOI: 10.3390/s20226458 PMID: 33198160

Dimitri, C., & Nehring, R. (2022). Thirty years of organic dairy in the United States: The influences of farms, the market and the organic regulation. *Renewable Agriculture and Food Systems*, *37*(6), 588–602. DOI: 10.1017/S1742170522000254

Džermeikaitė, K., Bačėninaitė, D., & Antanaitis, R. (2023). Innovations in Cattle Farming: Application of Innovative Technologies and Sensors in the Diagnosis of Diseases. *Animals (Basel)*, *13*(5), 780. Advance online publication. DOI: 10.3390/ani13050780 PMID: 36899637

Feng, H., Wang, X., Duan, Y., Zhang, J., & Zhang, X. (2020). Applying blockchain technology to improve agri-food traceability: A review of development methods, benefits and challenges. *Journal of Cleaner Production*, *260*, 121031. DOI: 10.1016/j.jclepro.2020.121031

Fidan, I., Huseynov, O., Ali, M. A., Alkunte, S., Rajeshirke, M., Gupta, A., Hasanov, S., Tantawi, K., Yasa, E., Yilmaz, O., Loy, J., Popov, V., & Sharma, A. (2023). Recent Inventions in Additive Manufacturing: Holistic Review. In *Inventions* (Vol. 8, Issue 4). DOI: 10.3390/inventions8040103

Fisher, O. J., Watson, N. J., Escrig, J. E., Witt, R., Porcu, L., Bacon, D., Rigley, M., & Gomes, R. L. (2020). Considerations, challenges and opportunities when developing data-driven models for process manufacturing systems. *Computers & Chemical Engineering*, *140*, 106881. DOI: 10.1016/j.compchemeng.2020.106881

Gehlot, A., Malik, P. K., Singh, R., Akram, S. V., & Alsuwian, T. (2022). Dairy 4.0: Intelligent Communication Ecosystem for the Cattle Animal Welfare with Blockchain and IoT Enabled Technologies. In *Applied Sciences* (Vol. 12, Issue 14). DOI: 10.3390/app12147316

Ghosal, A., Sagong, S., Halder, S., Sahabandu, K., Conti, M., Poovendran, R., & Bushnell, L. (2020). Truck platoon security: State-of-the-art and road ahead. *Computer Networks, 185*, 107658. DOI: 10.1016/j.comnet.2020.107658

Gowda, D. D., Sharma, A., Suneel, S., Kumar, N. A., Sharma, A., Suneel, S., & Manikandan, V. (2024). *Design and Implementation of an IoT-Based Control System for Precision Food Manufacturing*. DOI: 10.1007/978-981-97-5146-4_29

Groumpos, P. P. (2021). A Critical Historical and Scientific Overview of all Industrial Revolutions. *IFAC-PapersOnLine, 54*(13), 464–471. DOI: 10.1016/j.ifacol.2021.10.492 PMID: 38620687

Gupta, B. B., & Quamara, M. (2020). An overview of Internet of Things (IoT): Architectural aspects, challenges, and protocols. *Concurrency and Computation, 32*(21), e4946. DOI: 10.1002/cpe.4946

Hadi, M. S., Sugiono, B. S. R., Mizar, M. A., Witjoro, A., & Irvan, M. (2024). Enhancing low-temperature long-time milk pasteurization process with a C4.5 algorithm-driven AIoT system for real-time decision-making. *Journal of Food Process Engineering, 47*(4), e14606. DOI: 10.1111/jfpe.14606

Hassoun, A., Dankar, I., Bhat, Z., & Bouzembrak, Y. (2024). Unveiling the relationship between food unit operations and food industry 4.0: A short review. *Heliyon, 10*(20), e39388. DOI: 10.1016/j.heliyon.2024.e39388 PMID: 39492883

Hazra, A., Adhikari, M., Amgoth, T., & Srirama, S. (2021). A Comprehensive Survey on Interoperability for IIoT: Taxonomy, Standards, and Future Directions. *ACM Computing Surveys, 55*(1), 35. DOI: 10.1145/3485130

Heema, R., .S, S. & Gnanalakshmi, K. S. (2022). An Insight in to the Automation of the Dairy Industry: A Review. *Asian Journal of Dairy and Food Research*. DOI: 10.18805/ajdfr.DR-1856

Helo, P., & Hao, Y. (2021). Artificial intelligence in operations management and supply chain management: An exploratory case study. *Production Planning and Control, 33*(16), 1–18. DOI: 10.1080/09537287.2021.1882690

Hepfer, M., & Lawrence, T. (2022). The Heterogeneity of Organizational Resilience: Exploring functional, operational and strategic resilience. *Organization Theory, 3*(1). DOI: 10.1177/26317877221074701

Iftekhar, A., & Cui, X. (2021). Blockchain-Based Traceability System That Ensures Food Safety Measures to Protect Consumer Safety and COVID-19 Free Supply Chains. In *Foods* (Vol. 10, Issue 6). DOI: 10.3390/foods10061289

Iftekhar, A., Cui, X., Hassan, M., & Afzal, W. (2020). Application of Blockchain and Internet of Things to Ensure Tamper-Proof Data Availability for Food Safety. *Journal of Food Quality*, *2020*(1), 5385207. DOI: 10.1155/2020/5385207

Ivanov, D., Tang, C. S., Dolgui, A., Battini, D., & Das, A. (2021). Researchers' perspectives on Industry 4.0: Multi-disciplinary analysis and opportunities for operations management. *International Journal of Production Research*, *59*(7), 2055–2078. DOI: 10.1080/00207543.2020.1798035

Jadhav, A. R., R., S. K. M. P. & Pachamuthu, R. (. (2021). Development of a Novel IoT-Enabled Power- Monitoring Architecture With Real-Time Data Visualization for Use in Domestic and Industrial Scenarios. *IEEE Transactions on Instrumentation and Measurement*, *70*, 1–14. DOI: 10.1109/TIM.2020.3028437

Jerhamre, E., Carlberg, C. J. C., & van Zoest, V. (2022). Exploring the susceptibility of smart farming: Identified opportunities and challenges. *Smart Agricultural Technology*, *2*, 100026. DOI: 10.1016/j.atech.2021.100026

Kavitha, D., & Ravikumar, S. (2021). IOT and context-aware learning-based optimal neural network model for real-time health monitoring. *Transactions on Emerging Telecommunications Technologies*, *32*(1), e4132. DOI: 10.1002/ett.4132

Khanna, A., Jain, S., Burgio, A., Bolshev, V., & Panchenko, V. (2022). Blockchain-Enabled Supply Chain platform for Indian Dairy Industry: Safety and Traceability. *Foods*, *11*(17), 2716. DOI: 10.3390/foods11172716 PMID: 36076901

Kjønås, K., & Wangen, G. (2024). In Fritsch, L., Hassan, I., & Paintsil, E. (Eds.), *Cybersecurity Challenges and Smart Technology Adoption in Norwegian Livestock Farming BT - Secure IT Systems* (pp. 107–125). Springer Nature Switzerland.

Konstantinidis, F. K., Balaska, V., Symeonidis, S., Psarommatis, F., Psomoulis, A., Giakos, G., Mouroutsos, S. G., & Gasteratos, A. (2023). Achieving Zero Defected Products in Diary 4.0 using Digital Twin and Machine Vision. *Proceedings of the 16th International Conference on PErvasive Technologies Related to Assistive Environments*, 528–534. DOI: 10.1145/3594806.3596554

Kreuzberger, D., Kühl, N., & Hirschl, S. (2023). Machine Learning Operations (MLOps): Overview, Definition, and Architecture. *IEEE Access : Practical Innovations, Open Solutions*, *1*, 31866–31879. Advance online publication. DOI: 10.1109/ACCESS.2023.3262138

Kyaw, K. S., Adegoke, S. C., Ajani, C. K., Nwabor, O. F., & Onyeaka, H. (2024). Toward in-process technology-aided automation for enhanced microbial food safety and quality assurance in milk and beverages processing. *Critical Reviews in Food Science and Nutrition*, *64*(6), 1715–1735. DOI: 10.1080/10408398.2022.2118660 PMID: 36066463

Lal, P. P., Prakash, A. A., Chand, A. A., Prasad, K. A., Mehta, U., Assaf, M. H., Mani, F. S., & Mamun, K. A. (2022). IoT integrated fuzzy classification analysis for detecting adulterants in cow milk. *Sensing and Bio-Sensing Research*, *36*, 100486. DOI: 10.1016/j.sbsr.2022.100486

Leal, F., Chis, A. E., Caton, S., González–Vélez, H., García–Gómez, J. M., Durá, M., Sánchez–García, A., Sáez, C., Karageorgos, A., Gerogiannis, V. C., Xenakis, A., Lallas, E., Ntounas, T., Vasileiou, E., Mountzouris, G., Otti, B., Pucci, P., Papini, R., Cerrai, D., & Mier, M. (2021). Smart Pharmaceutical Manufacturing: Ensuring End-to-End Traceability and Data Integrity in Medicine Production. *Big Data Research, 24*, 100172. https://doi.org/https://doi.org/10.1016/j.bdr.2020.100172

Ledo, J., Hettinga, K. A., & Luning, P. A. (2020). A customized assessment tool to differentiate safety and hygiene control practices in emerging dairy chains. *Food Control, 111*, 107072. DOI: 10.1016/j.foodcont.2019.107072

Lee, E., Seo, Y.-D., Oh, S.-R., & Kim, Y.-G. (2021). A Survey on Standards for Interoperability and Security in the Internet of Things. *IEEE Communications Surveys and Tutorials*, *23*(2), 1020–1047. DOI: 10.1109/COMST.2021.3067354

Li, L., Rong, S., Wang, R., & Yu, S. (2021). Recent advances in artificial intelligence and machine learning for nonlinear relationship analysis and process control in drinking water treatment: A review. *Chemical Engineering Journal*, *405*, 126673. DOI: 10.1016/j.cej.2020.126673

Lohmer, J., & Lasch, R. (2020). Blockchain in operations management and manufacturing: Potential and barriers. *Computers & Industrial Engineering*, *149*, 106789. DOI: 10.1016/j.cie.2020.106789

Malik, M., Gahlawat, V. K., Mor, R. S., & Singh, M. K. (2024). Unlocking dairy traceability: Current trends, applications, and future opportunities. *Future Foods : a Dedicated Journal for Sustainability in Food Science*, *10*, 100426. DOI: 10.1016/j.fufo.2024.100426

Manivannan, D., Moni, S. S., & Zeadally, S. (2020). Secure authentication and privacy-preserving techniques in Vehicular Ad-hoc NETworks (VANETs). *Vehicular Communications*, *25*, 100247. DOI: 10.1016/j.vehcom.2020.100247

Marchi, B., Galati, C., & Zanoni, S. (2024). Management of spare parts for efficient maintenance: A case study in the dairy sector. *IFAC-PapersOnLine*, *58*(8), 389–394. DOI: 10.1016/j.ifacol.2024.08.152

Mehany, T., Siddiqui, S. A., Olawoye, B., Olabisi Popoola, O., Hassoun, A., Manzoor, M. F., & Punia Bangar, S. (2024). Recent innovations and emerging technological advances used to improve quality and process of plant-based milk analogs. *Critical Reviews in Food Science and Nutrition*, *64*(20), 7237–7267. DOI: 10.1080/10408398.2023.2183381 PMID: 36861223

Michie, C., Andonovic, I., Davison, C., Hamilton, A., Tachtatzis, C., Jonsson, N., Duthie, C.-A., Bowen, J., & Gilroy, M. (2020). The Internet of Things enhancing animal welfare and farm operational efficiency. *The Journal of Dairy Research*, *87*(S1), 20–27. DOI: 10.1017/S0022029920000680 PMID: 33213573

Morais, R., Mendes, J., Silva, R., Silva, N., Sousa, J. J., & Peres, E. (2021). A Versatile, Low-Power and Low-Cost IoT Device for Field Data Gathering in Precision Agriculture Practices. In *Agriculture* (Vol. 11, Issue 7). DOI: 10.3390/agriculture11070619

Moszyk, K., Deja, M., & Dobrzyński, M. (2021). Automation of the Road Gate Operations Process at the Container Terminal—A Case Study of DCT Gdańsk SA. *Sustainability (Basel)*, *13*(11), 6291. DOI: 10.3390/su13116291

Moudgil, V., Hewage, K., Hussain, S. A., & Sadiq, R. (2023). Integration of IoT in building energy infrastructure: A critical review on challenges and solutions. *Renewable & Sustainable Energy Reviews*, *174*, 113121. DOI: 10.1016/j.rser.2022.113121

Mrabet, H., Belguith, S., Alhomoud, A., & Jemai, A. (2020). A Survey of IoT Security Based on a Layered Architecture of Sensing and Data Analysis. In *Sensors* (Vol. 20, Issue 13). DOI: 10.3390/s20133625

Murphy, S. I., Reichler, S. J., Martin, N. H., Boor, K. J., & Wiedmann, M. (2021). Machine Learning and Advanced Statistical Modeling Can Identify Key Quality Management Practices That Affect Postpasteurization Contamination of Fluid Milk. *Journal of Food Protection*, *84*(9), 1496–1511. DOI: 10.4315/JFP-20-431 PMID: 33770185

Nagraik, R., Sharma, A., Kumar, D., Chawla, P., & Kumar, A. P. (2021). Milk adulterant detection: Conventional and biosensor based approaches: A review. *Sensing and Bio-Sensing Research*, *33*, 100433. DOI: 10.1016/j.sbsr.2021.100433

Najafi, M., Kaushal, V., & Visser, J. (2024). Operational Planning and Design Considerations for Underground Logistics Transportation in Texas. In *Infrastructures* (Vol. 9, Issue 8). DOI: 10.3390/infrastructures9080130

Naqvi, R., Soomro, T., Alzoubi, H., Ghazal, T. & Alshurideh, M. (2021). *The Nexus Between Big Data and Decision-Making: A Study of Big Data Techniques and Technologies*. DOI: 10.1007/978-3-030-76346-6_73

Neculai-Valeanu, A.-S., & Ariton, A.-M. (2022). Udder Health Monitoring for Prevention of Bovine Mastitis and Improvement of Milk Quality. *Bioengineering (Basel, Switzerland)*, *9*(11), 608. Advance online publication. DOI: 10.3390/bioengineering9110608 PMID: 36354519

Neethirajan, S. (2023). Artificial Intelligence and Sensor Technologies in Dairy Livestock Export: Charting a Digital Transformation. In *Sensors* (Vol. 23, Issue 16). DOI: 10.3390/s23167045

Newton, J. E., Nettle, R., & Pryce, J. E. (2020). Farming smarter with big data: Insights from the case of Australia's national dairy herd milk recording scheme. *Agricultural Systems*, *181*, 102811. DOI: 10.1016/j.agsy.2020.102811

Nikander, J., Manninen, O., & Laajalahti, M. (2020). Requirements for cybersecurity in agricultural communication networks. *Computers and Electronics in Agriculture*, *179*, 105776. DOI: 10.1016/j.compag.2020.105776

Nižetić, S., Šolić, P., López-de-Ipiña González-de-Artaza, D., & Patrono, L. (2020). Internet of Things (IoT): Opportunities, issues and challenges towards a smart and sustainable future. *Journal of Cleaner Production*, *274*, 122877. DOI: 10.1016/j.jclepro.2020.122877 PMID: 32834567

Novaes, A. G. N., Lima, O. F., De Cursi, J. E. S., Arias, J. A. C., & Santos, J. B. S. (2020). In Kotzab, H., & Pannek, J. (Eds.), *Predictive Manufacturing Tardiness Inference in OEM Milk-Run Operations BT - Dynamics in Logistics* (pp. 254–262). Springer International Publishing.

Nozari, H., Szmelter-Jarosz, A., & Ghahremani-Nahr, J. (2021). The Ideas of Sustainable and Green Marketing Based on the Internet of Everything—The Case of the Dairy Industry. In *Future Internet* (Vol. 13, Issue 10). DOI: 10.3390/fi13100266

Nyokabi, S. N., de Boer, I. J. M., Luning, P. A., Korir, L., Lindahl, J., Bett, B., & Oosting, S. J. (2021). Milk quality along dairy farming systems and associated value chains in Kenya: An analysis of composition, contamination and adulteration. *Food Control*, *119*, 107482. DOI: 10.1016/j.foodcont.2020.107482

Ordieres-Meré, J., Gutierrez, M., & Villalba-Díez, J. (2023). Toward the industry 5.0 paradigm: Increasing value creation through the robust integration of humans and machines. *Computers in Industry*, *150*, 103947. DOI: 10.1016/j.compind.2023.103947

Owusu-Kwarteng, J., Akabanda, F., Agyei, D., & Jespersen, L. (2020). Microbial Safety of Milk Production and Fermented Dairy Products in Africa. In *Microorganisms* (Vol. 8, Issue 5). DOI: 10.3390/microorganisms8050752

Pancholi, N., Gajera, H., & Shah, D. (2023). Improving quality of maintenance task for milk powder manufacturing unit through TOPSIS. *Journal of Quality in Maintenance Engineering*, *29*(1), 27–49. DOI: 10.1108/JQME-04-2021-0028

Pappas, N., Caputo, A., Pellegrini, M. M., Marzi, G., & Michopoulou, E. (2021). The complexity of decision-making processes and IoT adoption in accommodation SMEs. *Journal of Business Research*, *131*, 573–583. DOI: 10.1016/j.jbusres.2021.01.010

Pasupuleti, V., Thuraka, B., Kodete, C. S., & Malisetty, S. (2024). Enhancing Supply Chain Agility and Sustainability through Machine Learning: Optimization Techniques for Logistics and Inventory Management. In *Logistics* (Vol. 8, Issue 3). DOI: 10.3390/logistics8030073

Pu, Y.-Y., O'Donnell, C., Tobin, J. T., & O'Shea, N. (2020). Review of near-infrared spectroscopy as a process analytical technology for real-time product monitoring in dairy processing. *International Dairy Journal*, *103*, 104623. DOI: 10.1016/j.idairyj.2019.104623

Rathakrishnan, T., Suntharalingam, C., Yi Shern, L., Loong, S. K., Lee, H., Tee, K., Thanarajoo, S., & Balasundram, S. (2022). Adoption of Sustainable Agricultural Practices Among Smallholder Dairy Farmers in Malaysia: Contributing Factors and Smart Farming Prospects. *Journal of Smart Science and Technology*, *2*(2), 16–31. DOI: 10.24191/jsst.v2i2.34

Rejeb, A., Keogh, J., Zailani, S., Treiblmaier, H., & Rejeb, K. (2020). Blockchain Technology in the Food Industry: A Review of Potentials, Challenges and Future Research Directions. *Logistics (Basel)*, *4*(4), 27. DOI: 10.3390/logistics4040027

Sahoo, S., & Lo, C.-Y. (2022). Smart manufacturing powered by recent technological advancements: A review. *Journal of Manufacturing Systems*, *64*, 236–250. DOI: 10.1016/j.jmsy.2022.06.008

Sar, T., Harirchi, S., Ramezani, M., Bulkan, G., Akbas, M. Y., Pandey, A., & Taherzadeh, M. J. (2022). Potential utilization of dairy industries by-products and wastes through microbial processes: A critical review. *The Science of the Total Environment*, *810*, 152253. DOI: 10.1016/j.scitotenv.2021.152253 PMID: 34902412

Schiller, E., Aidoo, A., Fuhrer, J., Stahl, J., Ziörjen, M., & Stiller, B. (2022). Landscape of IoT security. *Computer Science Review*, *44*, 100467. DOI: 10.1016/j.cosrev.2022.100467

Shahin, M., Chen, F. F., Hosseinzadeh, A., & Zand, N. (2023). Using machine learning and deep learning algorithms for downtime minimization in manufacturing systems: An early failure detection diagnostic service. *International Journal of Advanced Manufacturing Technology*, *128*(9), 3857–3883. DOI: 10.1007/s00170-023-12020-w

Sinha, B. B., & Dhanalakshmi, R. (2022). Recent advancements and challenges of Internet of Things in smart agriculture: A survey. *Future Generation Computer Systems*, *126*, 169–184. DOI: 10.1016/j.future.2021.08.006

Soon, J. M., & Abdul Wahab, I. R. (2021). Global food recalls and alerts associated with labelling errors and its contributory factors. *Trends in Food Science & Technology*, *118*, 791–798. DOI: 10.1016/j.tifs.2021.11.001

Soon, J. M., Brazier, A. K. M., & Wallace, C. A. (2020). Determining common contributory factors in food safety incidents – A review of global outbreaks and recalls 2008–2018. *Trends in Food Science & Technology*, *97*, 76–87. DOI: 10.1016/j.tifs.2019.12.030

Sowmya, N., & Ponnusamy, V. (2021). Development of Spectroscopic Sensor System for an IoT Application of Adulteration Identification on Milk Using Machine Learning. *IEEE Access : Practical Innovations, Open Solutions*, *9*, 53979–53995. DOI: 10.1109/ACCESS.2021.3070558

Syrmos, E., Bechtsis, D., Vlachos, D., Papapanagiotakis, G., Todi, T., Papaspyropoulou, M., Georgakidis, K., & Sfitis, N. (2023). In Rabie, K., Kharel, R., Mohjazi, L., Elganimi, T., & Khan, W. U. (Eds.), *An IoT Monitoring System for Dairy Products BT - IoT as a Service* (pp. 12–22). Springer Nature Switzerland.

Taheri, R., Shojafar, M., Alazab, M., & Tafazolli, R. (2021). Fed-IIoT: A Robust Federated Malware Detection Architecture in Industrial IoT. *IEEE Transactions on Industrial Informatics*, *17*(12), 8442–8452. DOI: 10.1109/TII.2020.3043458

Tan, A., & Ngan, P. T. (2020). A proposed framework model for dairy supply chain traceability. *Sustainable Futures : An Applied Journal of Technology, Environment and Society*, *2*, 100034. DOI: 10.1016/j.sftr.2020.100034

Theissler, A., Pérez-Velázquez, J., Kettelgerdes, M., & Elger, G. (2021). Predictive maintenance enabled by machine learning: Use cases and challenges in the automotive industry. *Reliability Engineering & System Safety*, *215*, 107864. DOI: 10.1016/j.ress.2021.107864

Theodorakopoulos, L., Theodoropoulou, A., & Stamatiou, Y. (2024). A State-of-the-Art Review in Big Data Management Engineering: Real-Life Case Studies, Challenges, and Future Research Directions. In *Eng* (Vol. 5, Issue 3, pp. 1266–1297). DOI: 10.3390/eng5030068

Trakadas, P., Simoens, P., Gkonis, P., Sarakis, L., Angelopoulos, A., Ramallo-González, A. P., Skarmeta, A., Trochoutsos, C., Calvo, D., Pariente, T., Chintamani, K., Fernandez, I., Irigaray, A. A., Parreira, J. X., Petrali, P., Leligou, N., & Karkazis, P. (2020). An Artificial Intelligence-Based Collaboration Approach in Industrial IoT Manufacturing: Key Concepts, Architectural Extensions and Potential Applications. *Sensors (Basel)*, *20*(19), 5480. Advance online publication. DOI: 10.3390/s20195480 PMID: 32987911

Uzoigwe, D. (2024). Evaluating the Effectiveness of Reliability-Centered Maintenance Programs in Food and Beverage Manufacturing Facilities; A Review. *International Journal of Research and Innovation in Applied Science*, *9*(2), 204–232. Advance online publication. DOI: 10.51584/IJRIAS.2024.90218

Varma, V. S., Parajuli, R., Scott, E., Canter, T., Lim, T. T., Popp, J., & Thoma, G. (2021). Dairy and swine manure management – Challenges and perspectives for sustainable treatment technology. *The Science of the Total Environment*, *778*, 146319. DOI: 10.1016/j.scitotenv.2021.146319 PMID: 33721638

Vejan, P., Khadiran, T., Abdullah, R., & Ahmad, N. (2021). Controlled release fertilizer: A review on developments, applications and potential in agriculture. *Journal of Controlled Release*, *339*, 321–334. DOI: 10.1016/j.jconrel.2021.10.003 PMID: 34626724

Venkatesh, V. G., Kang, K., Wang, B., Zhong, R. Y., & Zhang, A. (2020). System architecture for blockchain based transparency of supply chain social sustainability. *Robotics and Computer-integrated Manufacturing*, *63*, 101896. DOI: 10.1016/j.rcim.2019.101896

Wangen, S. R., Zhang, F., Fadul-Pacheco, L., da Silva, T. E., & Cabrera, V. E. (2021). Improving farm decisions: The application of data engineering techniques to manage data streams from contemporary dairy operations. *Livestock Science*, *250*, 104602. DOI: 10.1016/j.livsci.2021.104602

Wu, J.-Y., & Hsiao, H.-I. (2021). Food quality and safety risk diagnosis in the food cold chain through failure mode and effect analysis. *Food Control*, *120*, 107501. DOI: 10.1016/j.foodcont.2020.107501

Ziyaina, M., Rasco, B., & Sablani, S. S. (2020). Rapid methods of microbial detection in dairy products. *Food Control*, *110*, 107008. DOI: 10.1016/j.foodcont.2019.107008

KEY TERMS AND DEFINITIONS

Data Analytics: The systematic computational analysis of data collected via IoT devices to identify patterns, trends, and actionable insights for process optimization and decision-making.

Industry 4.0: The current trend of automation and data exchange in manufacturing technologies, integrating cyber-physical systems, IoT, and cloud computing for smart industrial operations.

Internet of Things (IoT): A network of interconnected physical devices embedded with sensors, software, and other technologies to collect and exchange data over the internet in real time.

Machine Learning: A branch of artificial intelligence (AI) that uses algorithms to learn from data and improve predictive capabilities without explicit programming, commonly used for predictive maintenance and quality forecasting.

Operational Efficiency: The capability to deliver products or services cost-effectively while maintaining quality and productivity, achieved through real-time monitoring and automation in IoT-enabled facilities.

Predictive Maintenance: A proactive maintenance strategy that uses IoT data and machine learning models to predict equipment failures before they occur, reducing downtime and repair costs.

Quality Control: The process of ensuring dairy products meet safety and quality standards by monitoring critical control points using IoT technologies in real time.

Sensors and Actuators: Sensors detect and measure physical properties (e.g., temperature, humidity) while actuators convert control signals into physical actions (e.g., opening valves), enabling real-time process control.

Traceability: The ability to track the history, application, or location of dairy products throughout the supply chain, enhanced by IoT for improved transparency and food safety compliance.

Chapter 6
IoT-Driven Automation in Dairy Processing for Waste Reduction and Productivity Enhancement

Leena Arya
https://orcid.org/0000-0002-9249-732X
Koneru Lakshmaiah Education Foundation, India

Ravi Rastogi
https://orcid.org/0000-0002-1118-3020
Koneru Lakshmaiah Education Foundation, India

Monika Bhatnagar
https://orcid.org/0000-0003-0005-882X
Galgotias College of Engineering and Technology, India

Latha Banda
https://orcid.org/0000-0003-1189-1569
JIMS Engineering Management Technical Campus, Greater Noida, India

Shashi Mehrotra
https://orcid.org/0000-0003-3907-5790
SRM Institute of Science and Technology, India

Devendra Gautam
https://orcid.org/0009-0005-3666-3257
JIMS Engineering Management Technical Campus, Greater Noida, India

Nagamalli Arasavalli
https://orcid.org/0000-0002-5002-5615
Koneru Lakshmaiah Education Foundation, India

ABSTRACT

The dairy industry is vital in the global agricultural economy, providing essential nutrition and significant economic contributions. However, it faces persistent challenges such as inefficiencies, waste, and environmental unsustainability. Traditional

DOI: 10.4018/979-8-3693-9949-1.ch006

dairy processing systems, characterised by high resource consumption and limited technological integration, need help to meet modern requirements for efficiency and sustainability. The internet of things (IoT) has become a solution that offers real-time monitoring, predictive maintenance, process optimization, and traceability of supply chains. This chapter addresses IoT-driven automation in dairy processing, emphasizing the reduction of waste, increased productivity, and quality consistency. It discusses challenges in adopting IoT, including initial high costs, interfacing with legacy systems, data security issues, and strategies for overcoming these barriers. Through IoT-enabled systems, the dairy industry will be able to move toward a sustainable and connected future, meeting global demand with minimal environmental footprint.

INTRODUCTION

Dairy industries play a key role in the world agricultural economy, supporting food security and nutrition through considerable contributions to human development and economic development. According to Food and Agriculture Organization (FAO) statistics, up to 2023, global milk production exceeds 930 million metric tons, and population growth, urbanization, and growing consumer demand are driving these levels upward (FAO, 2023). Despite its significance, the industry is characterized by persistent inefficiency, waste, and lack of environmental sustainability. These are increasingly significant issues as global consumption patterns change toward higher demand for processed dairy products, which requires more sophisticated and sustainable production systems (Kumar & Joshi, 2021).

Among the most critical challenges facing dairy processing is waste management. Studies reveal an annual wastage of about 20% of dairy products across the supply chain, translating into millions of metric tons of milk and its derivatives (Smith & Wang, 2020). Apart from being an economic loss, wages increase environmental problems such as more greenhouse gas emissions and the consumption of resources. Dairy processing is considered one of the most water- and energy-consuming industries. For example, it takes 4–7 litres of water to produce one litre of milk. Energy-intensive processes like pasteurization, homogenization, and packaging also increase the dairy industry's carbon footprint (Libelium, n.d.; Gupta & Roy, 2023). In response to the growing need for sustainability, the dairy industry is increasingly under pressure from regulatory bodies to make the business cleaner, more efficient, and less wasteful.

The Internet of Things (IoT) has opened new avenues to address these challenges. IoT integrates physical devices, sensors, and data analytics into a connected ecosystem capable of real-time monitoring, decision-making, and process optimization. IoT offers transformative potential across the entire value chain in dairy processing, from raw milk procurement and quality monitoring to waste management and product distribution. For instance, IoT sensors in milk collection systems can continually monitor parameters such as temperature, pH, and bacterial load, thus ensuring that only good-quality milk goes into the processing chain (Lee & Lee, 2015). IoT-based automation in processing facilities enables dynamic adjustment of production parameters, reducing material waste, boosting energy efficiency, and guaranteeing improved product consistency (Fitcows, n.d.).

IoT also makes predictive maintenance possible, which extends the equipment's lifespan and decreases downtime from malfunctions. IoT-based smart waste management systems track real-time waste generation in dairy plants and help identify inefficiencies. These systems further optimize water and energy consumption through recycling and reuse, reducing the environmental footprint in dairy processing (Arya et al., 2023a; Arya et al., 2023b). When integrated with advanced data analytics and AI, IoT's actionable insights further enhance productivity and sustainability in decision-making (Libelium, n.d.; Mhapsekar et al., 2024).

The global dairy industry is estimated at approximately $500 billion in 2022 and is forecast to grow at a compound annual growth rate (CAGR) of 5% through 2030. To match the demand for this sector while responding to environmental and economic pressure, adopting IoT-driven automation becomes imperative, not an option. Improving efficiency in operations, cutting resource waste, and ensuring quality products through IoT can change the fabric of traditional dairy processing and position the industry towards a sustainable future (Lee & Lee, 2015; Arya et al., 2023a).

This chapter explores the transformative capacity of IoT-driven automation in dairy processing relating to waste reduction and productivity enhancement. It discusses IoT applications at key stages in dairy production, introduces real-world case studies of successful implementations, and examines challenges associated with IoT adoption. Finally, the chapter explores future directions for research and innovation, providing a roadmap for stakeholders in academia, industry, and policy-making to leverage IoT for a more sustainable and efficient dairy industry.

THE DAIRY INDUSTRY: AN OVERVIEW

The milk industry is one of the largest sectors of the global food economy. It provides important nutrition to billions of humans worldwide and supports millions of farmers and their families. It stretches from production and collection to processing and distribution, conveying products such as milk, cheese, yoghurt, butter, and powdered milk. The industry plays a significant role in world nutrition, providing top-quality protein, calcium, and essential vitamins to fulfil millions' dietary needs. According to FAO, dairy production represents around 10% of world agricultural output, with milk ranking third among the most-produced agricultural commodities after rice and wheat (FAO, 2023).

Global milk production exceeded 930 million metric tons in 2023, and its demand is expected to continue growing steadily with the rising population, urbanization, and per capita consumption. The top milk-producing countries include India, the United States, China, and Brazil, with India accounting for about 23% of global milk output (Zarin et al., 2023). Despite its strategic importance, the sector is currently challenged by inefficient resource utilization, environmental effects, and sustainability requirements, which are maintained with increasing global demand (Ahmad et al., 2019).

The dairy industry offers massive economic contributions to agriculture; it is the foundational tool for rural development and jobs. Smallholder dairy producers significantly contribute to milk production volumes, especially in developing regions. On the other hand, milk production in industrial countries involves large-scale farms and modern processing plants; there is a domestic demand for and export of such products. Gross Domestic Products in many countries benefit hugely from the dairy industry while anchoring food security and economic stability.

Dairy Industry Structure

The dairy industry is a complex value chain involving several stages and stakeholders. It can be broadly categorised into two primary segments: Production, processing and Distribution.

Primary Production

This stage includes the production of raw milk at the farm level. Key activities include:

- Dairy Farming: Milk is collected from dairy animals, including cows, buffaloes, goats, and sheep. High yields and good-quality milk depend mainly on feeding, breeding, and proper livestock health management (Galukande et al., 2013).
- Milk Collection: Bulk tanks store milk mainly at rural collection centers or cooperative societies. Quality tests include percentage fat content, pH levels, and bacterial count at collection time.
- Transportation: Milk is transported to processing units under controlled conditions to prevent spoilage, usually in refrigerated tankers (Zarin et al., 2023).

Primary production differs significantly from region to region. Smallholder farmers dominate developing countries, whereas large-scale, automated dairy farms are more common in developed countries. Modern farms use automated milking systems, GPS cattle tracking, and feed optimization for improved efficiency.

Processing and Distribution

This stage processes raw milk into consumable products and distributes them to the final consumers. It includes:

Processing:

- Pasteurization: Heating milk to kill harmful microorganisms and increase shelf life.
- Homogenization: Breaking down fat globules makes the texture uniform and increases stability.
- Fermentation: Making yoghurt, cheese, and other cultured dairy products.
- Drying and Condensing: Produce powders and condensed milk for long-term storage and export.
- Packaging: Advanced packaging techniques ensure product safety and increase shelf life.

Distribution:

- Distribution to consumers via retail chains, supermarkets, and direct-to-consumer channels.
- Cold chain logistics for transporting and storing products (Aung & Chang, 2014).

Key Players in the Dairy Industr

The dairy value chain is made up of several stakeholders with a broad variety of interests:

- Farmers: Producing milk is the initial stage of quality control.
- Cooperatives: The most important organizations for gathering milk, guaranteeing its quality, and providing a market for small-scale farmers are cooperatives.
- Processing Companies: Use cutting-edge technologies to turn raw milk into a variety of dairy products.
- Retailers and Distributors: Make sure processed goods are delivered to customers while upholding quality requirements.

- o Consumers: Drive demand for diverse, high-quality, sustainable dairy products.

CHALLENGES IN TRADITIONAL DAIRY PROCESSING

Despite serving as the foundation of the dairy business for many years, traditional processing techniques are progressively falling short of meeting the contemporary demands for quality, sustainability, and efficiency. Traditional systems have numerous obstacles that reduce productivity, raise waste, and affect environmental sustainability as the demand for dairy products rises globally. The main difficulties in conventional dairy processing are listed below:

Resource Intensity

- o Excessive Water Use: Water is essential to the cleansing, chilling, and processing processes of traditional dairy production. Four to seven litres of water are thought to be needed to digest one litre of milk.
- o This excessive consumption strains freshwater resources, particularly in regions facing water scarcity (FAO, 2023; Ahmad et al., 2019).
- o Energy-Intensive Processes: Pasteurisation, homogenisation, and refrigeration consume substantial energy. Traditional systems' dependence on non-renewable energy sources further increases their environmental footprint (Gupta & Roy, 2023).

Waste Generation

- o Solid Waste: Dairy processing generates significant amounts of solid waste, including spoiled milk, by-products, and packaging materials. Improper management of these wastes contributes to pollution and economic losses (Shamsuddoha & Nasir, 2024).
- o Liquid Waste: Effluents from cleaning, washing, and processing contain high levels of organic matter, fats, and nutrients. If discharged untreated, these can cause environmental damage, contributing to water pollution and eutrophication (Ahmad et al., 2019).

Safety and Quality Concerns

- o Product Quality Variability: Product quality is often inconsistent because of the inability of conventional systems to monitor and control critical param-

- eters such as pH, temperature, and microbial activity in real-time (Fitcows, n.d.).
- Risks to the Safety of Food: Inadequate monitoring raises the danger of contamination, spoilage, and noncompliance with food safety regulations. Product recalls and financial losses could result from declining customer trust (Dabbene et al., 2014; Aung & Chang, 2014).

Inefficient Supply Chains

- Milk Spoilage: Raw milk frequently spoils due to ineffective and delayed transportation from farms to processing plants, especially in areas with disjointed supply chains and insufficient cold storage facilities.
- Traceability Challenges: Because traditional systems cannot follow products from beginning to finish, it is difficult to find supply chain issues or verify item legitimacy.

Equipment Downtime

- Unplanned Maintenance: Traditional equipment often lacks predictive maintenance capabilities, resulting in frequent breakdowns and unplanned downtime. This disrupts production schedules and increases operational costs (Mhapsekar et al., 2024).
- Aging Infrastructure: Many dairy processing plants rely on outdated machinery that is less efficient and more prone to failures, further hindering productivity (Nestlé Sustainability Reports, 2021).

Environmental Impact

- Greenhouse Gas Emissions: The dairy processing industry contributes significantly to greenhouse gas emissions through energy use, wastewater management, and transportation (Libelium, n.d.). Traditional practices often lack mechanisms to reduce or offset these emissions.
- Excessive By-Products: Whey and other by-products are frequently thrown away or used insufficiently in the absence of sophisticated utilisation procedures, which leads to resource waste (Ahmad et al., 2019).

Labor Dependency

- o Manual Operations: Traditional dairy processing heavily depends on manual labour for monitoring, controlling, and adjusting processes. This limits scalability and increases the risk of human error (Mahalik & Nambiar, 2010).
- o Workforce Limitations: Labor shortages, particularly in rural areas, can disrupt operations, and reliance on unskilled labour can compromise quality and efficiency (Galukande et al., 2013).

Inadequate Data Use

- o Lack of Real-Time Monitoring: Manual inspections that are infrequent, imprecise, and time-consuming are often the foundation of conventional systems. The ability to respond quickly to equipment malfunctions or inconsistent goods is so limited.
- o Inefficient Data Use: Operators are unable to learn about performance patterns, resource consumption, and areas for development when there are no integrated data systems in place (Kumar & Joshi, 2021).

Regulatory and Market Pressures

- o Increasing Compliance Requirements: As global food safety and environmental regulations become more stringent, traditional systems struggle to adapt due to limited monitoring and reporting capabilities (Dabbene et al., 2014).
- o Growing Consumer Expectations: Modern consumers demand high-quality, sustainably produced dairy products. Traditional methods often fail to meet these expectations, resulting in competitive disadvantages (Lee & Lee, 2015).

THE ROLE OF TECHNOLOGY IN TRANSFORMING THE DAIRY INDUSTRY

Technological advancements in dairy are changing the scenario for these challenges. The Internet of Things (IoT) encourages smart and sustainable dairy processing. IoT allows real-time monitoring of production parameters, prediction of equipment maintenance, optimization of resources, which directly reduces waste and enhances productivity (Mhapsekar et al., 2024; Libelium, n.d.). Automated systems of dairy processing ensure consistency in quality.

Furthermore, IoT applications reach the primary production stage through smart farming solutions such as automated milking robots, GPS-enabled cattle tracking, and feed optimization systems that transform dairy farming practices. These technologies improve animal health, milk yield, and operational efficiency, ensuring a steady supply of high-quality milk to processors.

The Internet of Things (IoT) revolutionises the dairy processing industry with real-time monitoring, predictive maintenance, and process optimization. This technology improves efficiency, reduces waste, and provides quality products (Nagraj et al., 2022; Almaiah et al., 2023). Figure 1. represents the IoT-driven automatic dairy processing system.

Figure 1. IoT-driven automatic dairy processing

Real-Time Monitoring of Milk Quality

IoT technologies ensure that quality monitoring is consistent across all stages of dairy processing. Real-time sensors monitor critical parameters such as temperature, pH, and microbial counts for product safety and regulatory compliance (FAO, 2023; Fitcows, n.d.). The Food and Agriculture Organization estimates that global milk spoilage rates average 20% (FAO, 2023). IoT integration has been shown to reduce spoilage by 15-20%, mainly through optimized storage and transportation systems. Table 1. represents the key parameters monitored in IoT-driven dairy processing system.

IoT sensors make it possible to monitor important aspects of milk quality in real time, guaranteeing uniformity and safety throughout the whole processing process.

- **Temperature Control**: To prevent spoiling, milk must always be maintained at the right temperatures. Because IoT-based temperature sensors give real-time data, if the temperature changes, corrective action may be done right away. Implementing IoT-based temperature monitoring systems has reduced spoilage rates by 20% (Fitcows, n.d.; Libelium, n.d.).
- **pH Levels**: Checking pH levels reveals contamination and ensures product uniformity. IoT sensors allow continuous pH measuring, thus allowing a timely response to any adverse changes (Smith & Wang, 2020).
- **Fat and Protein Content**: Checking milk's nutritional ingredients is important for quality monitoring. IoT devices can measure fat and protein content in real-time and ensure adherence to product specifications (Gupta & Roy, 2023).

Table 1. Key parameters monitored in IoT-driven dairy processing

Parameter	Optimal Range	IoT Application	Improvement in Quality (%)
Temperature	2°C – 4°C	Real-time cooling system monitoring	18%
pH Level	6.5 – 6.8	Automated pH sensors	15%
Microbial Count	< 50,000 CFU/mL	Microbial detection systems	20%

Predictive Maintenance of Equipment

IoT facilitates predictive maintenance by monitoring equipment health indicators and minimizing downtime and maintenance costs (Amul Dairy Cooperative, 2022; Mhapsekar et al., 2024). Figure 2. shows how IoT-based predictive maintenance can reduce equipment downtime.

- o Vibration Analysis: Sensors identify unusual vibrations in equipment, which may be a sign of failure. Costly malfunctions are avoided by timely maintenance made possible by early identification.
- o Temperature Monitoring: Increases in temperature in equipment parts can indicate wear or malfunction. IoT devices track such changes, allowing preemptive intervention (Fitcows, n.d.; Kumar & Joshi, 2021).
- o Energy Consumption: Energy usage patterns indicate inefficiencies. IoT systems alert operators to anomalies, and energy-saving interventions can be undertaken (Nestlé Sustainability Reports, 2021).

Figure 2. Reduction in equipment downtime with IoT-based predictive maintenance

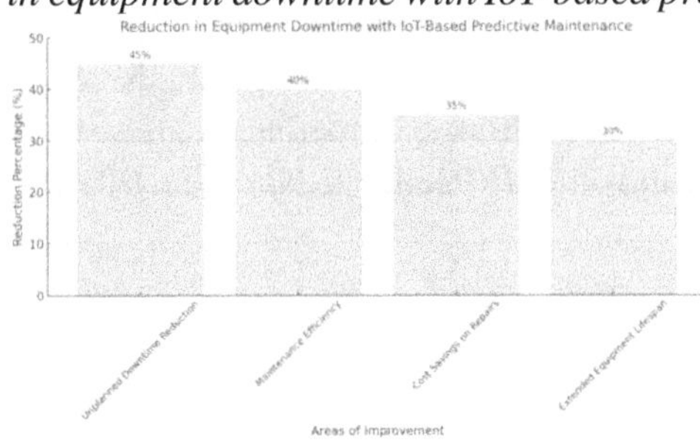

Process Optimization

With real-time data and control, IoT has the potential for dairy processing activities to be optimized.

- o Pasteurization Control: In order to guarantee product safety during pasteurization, temperature control is essential. Precision temperature management is made possible by IoT systems, increasing its effectiveness and reliability (Gupta & Roy, 2023).

- o Homogenization Efficiency: In homogenization, the pressure and flow rate should be maintained constant to ensure uniform texture in the product. IoT devices help adjust for optimal conditions (Mahalik & Nambiar, 2010).
- o Fermentation Monitoring: In yogurt and cheese manufacturing, IoT sensors monitor acidity and temperature to achieve desired product characteristics (Ahmad et al., 2019).

Smart Feeding Technology

Real-time monitoring of livestock feed intake, nutritional value, and waste is possible using Internet of Things-based smart feeding systems. This reduces feed waste and guarantees the animals receive the best nutrition possible. Figure 3 illustrates how smart feeding technology incorporates sensor data from automated feeders and GPS trackers to dynamically modify feed amounts according to the health and productivity requirements of the animal. Among the main advantages of cutting-edge feeding technology are:

- o Improved Feed Efficiency: Reduces overfeeding and underfeeding, optimizing the feed-to-milk conversion ratio.
- o Cost Savings: Reduces feed wastage and lowers overall feeding costs by up to 10–15% (Gupta & Roy, 2023; Galukande et al., 2013).
- o Environmental Benefits: Minimizes methane emissions from livestock by optimizing feed composition (Mahalik & Nambiar, 2010).

Figure 3. Smart feeding technology

Waste Reduction and Resource Management

IoT in dairy processing significantly reduces waste and conserves resources. Table 2. shows the impact of IoT on resource utilization.

- Water Consumption Monitoring: IoT sensors measure water consumption in real-time, thus identifying where it could be saved. This has reduced water consumption by 15% in some dairy plants (Fitcows, n.d.; FAO, 2023).
- Energy Management: IoT systems analyze energy consumption patterns to optimize usage, save costs, and reduce negative environmental impacts.
- By-Product Utilization: Process monitoring can identify valuable by-products for efficient use or recycling (Shamsuddoha & Nasir, 2024).

Table 2. Impact of IoT on resource utilization

Resource	Traditional Usage	IoT-Enabled Usage	Reduction (%)
Water	7 litres /litre of milk	5.5 litres/litre	20%
Energy	15 kWh/ton	12.5 kWh/ton	15%
Raw Material	1.2 kg/litre of product	1.1 kg/litre	8%
Feed Wastage	25%	10%	15%

Supply Chain Traceability

IoT Traceability across the dairy supply chain ensures high quality and safety from farm to consumer (Aung & Chang, 2014; Dabbene et al., 2014). Figure 4. represents the improvement in supply chain transparency with IoT implementation.

- o RFID Tagging: RFID tags attached to the milk can be tracked throughout the whole supply chain, thus enabling transparency and accountability (Gupta & Roy, 2023; IoT ONE, n.d.).
- o Blockchain Integration: By integrating the IoT with the blockchain network, data is secured for an immutable record of the product's movement (Mhapsekar et al., 2024).
- o Quality Assurance: Real-time information about storage and transit conditions ensures that product quality is maintained, and spoilage is not at risk (Libelium, n.d.).

Figure 4. Improvement in supply chain transparency with IoT implementation

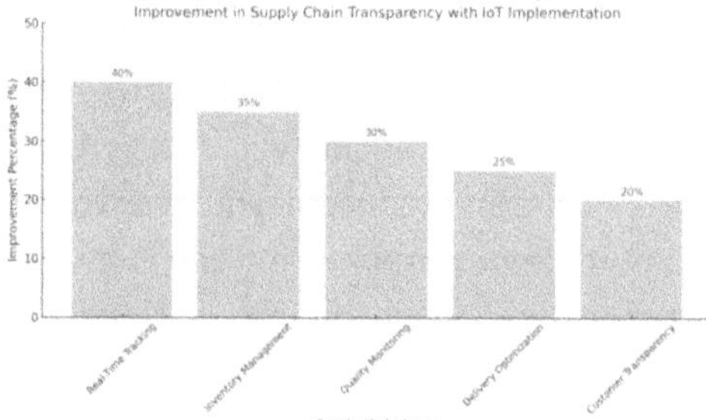

PROPOSED METHODOLOGY

The process begins with initiating IoT-enabled monitoring and automation systems in the dairy plant. This will provide seamless integration of sensors, devices, and analytics platforms. Figure 5. depicts the flowchart of the process of IoT-enabled monitoring and automation systems in the dairy plant.

Figure 5. Flowchart of the process of IoT-enabled monitoring and automation systems in the dairy plant

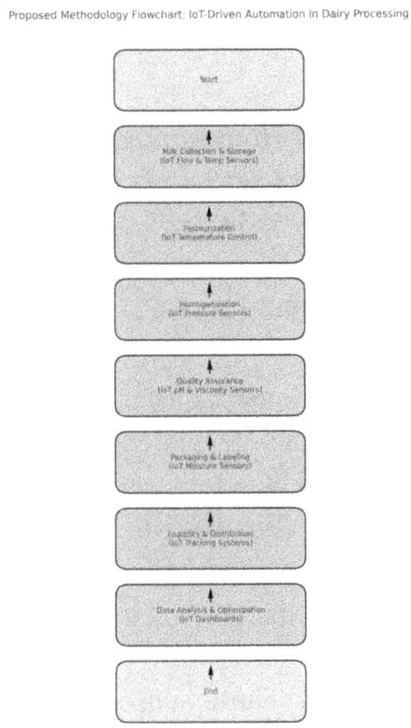

The flowchart outlines the steps in IoT-driven automation in dairy processing while focusing on waste reduction and productivity enhancement:

Collection & Storage

- **IoT Flow and Temperature Sensors**: IoT flow sensors track the incoming volume during milk collection to avoid spillage or overflows. Temperature sensors ensure milk stays fresh under optimal conditions (Arya et al., 2023a; Kumar & Joshi, 2021).

- o **Real-Time Monitoring**: Sensor data is transmitted to a centralised system, ensuring immediate corrective actions if anomalies, such as temperature fluctuations, are detected (Fitcows, n.d.; FAO, 2023).

Pasteurization

- o **IoT Temperature Control**: Pasteurization is carried out on the milk to kill harmful bacteria yet preserve nutrients. IoT sensors precisely monitor and control this process, thereby maintaining quality while adhering to food safety requirements (Smith & Wang, 2020).
- o **Automation**: IoT-enabled systems can automate heating cycles, allowing optimal energy use and wastage reduction (Nestlé Sustainability Reports, 2021).

Homogenization

- o **IoT Pressure Sensors**: Homogenization breaks fat molecules into a uniform texture. Pressure sensors help the process to run at optimum levels and avoid under or over-processing (Mahalik & Nambiar, 2010).
- o **Efficiency**: Real-time adjustments do not waste resources and ensure product consistency (Amul Dairy Cooperative, 2022).

Quality Assurance

- o **IoT pH and Viscosity Sensors**: The quality of milk or processed dairy products is assessed at this stage. Sensors with IoT capabilities monitor the pH levels, viscosity, etc., to ensure these products meet the quality criteria (Libelium, n.d.; Gupta & Roy, 2023).
- o **Automated Feedback Loops**: Any deviation sensed by the sensors gives an alert or adjusts the product to avert human error.

Packaging & Labeling

- o **IoT Moisture Sensors**: These sensors make sure that the product is clean and that the packaging materials are appropriate. Accuracy and traceability are guaranteed by the automated labeling procedure.
- o **Waste Reduction**: Automated packaging boosts efficiency and uses less material.

Distribution & Logistics

- o **IoT Tracking Systems**: IoT-enabled tracking keeps an eye on temperature and humidity during transit to ensure the integrity of the product once it is ready for use (IoT ONE, n.d.; Aung & Chang, 2014).
- o **Supply Chain Optimization**: Real-time position tracking and updates are made possible by IoT devices, which increases the efficiency and transparency of the supply chain (Dabbene et al., 2014).

Data Analysis & Optimization

- o **IoT Dashboards**: All collected data is aggregated and analysed on IoT dashboards. Insights from this data drive process optimisation, such as improving resource allocation, reducing energy consumption, and identifying bottlenecks (Gupta & Roy, 2023; Mhapsekar et al., 2024).

- **Predictive Analytics**: By anticipating maintenance requirements and process enhancements, the system guarantees ongoing productivity.

The process ends with a productive, waste-free operation where IoT technologies have optimized productivity, lowered operating costs, and guaranteed constant quality. Through data-driven optimization, automation, and real-time monitoring, this structured technique shows how IoT-enabled automation enhances dairy processing. It meets the demands of contemporary industry by ensuring sustainability and minimizing waste.

CASE STUDIES OF IOT IMPLEMENTATION IN DAIRY PROCESSING

The case study discusses how IoT innovations have transformed the operational setup of dairy industries worldwide through downtime, resource optimisation, milk production, and general efficiency.

Amul, India, implemented IoT-based predictive maintenance systems and automated quality monitoring to enhance equipment efficiency and reduce downtimes. The outcomes were remarkable, with a 25% reduction in downtime, 30% savings in maintenance costs, and a 20% improvement in milk quality consistency (Amul Dairy Cooperative, 2022). Nestlé, Europe focused on sustainability by deploying IoT-enabled sensors for water recycling, energy optimization, and by-product utilization. This strategy resulted in considerable resource optimization, reduced water usage by 20%, increased energy consumption by 15%, and successfully repurposed whey into high-value protein supplements (Nestlé Sustainability Reports, 2021). At Voshazhnikovo Farm, Russia, IoT sensors were used to monitor environmental conditions and cow behaviour. Integrated with machine learning, the system gave an 18% rise in milk production and €340,000 in cost savings over 180 days on feed, indicating that technology in dairy farming drastically changes productivity (Libelium, n.d.). Stellapps Technologies, India, introduced IoT applications to improve productivity and profitability in dairy farming. Real-time monitoring and managing dairy operations led to measurable improvements in operational efficiency and financial outcomes (IoT ONE, n.d.). The Edge-Centric Industrial IoT Solution demonstrated the power high technologies such as Edge-AI and smart sensors can bring to optimise dairy processing activities. It featured real-time monitoring and control capabilities, dramatically improving productivity and efficiency (Mhapsekar et al., 2024). Table 3 presents case study examples of operational applications in dairy processing using IoT-driven automation and their objectives, implementation, and outcomes.

Table 3. Case studies of operational applications in dairy processing using IoT-driven automation

Case Study	Objective	Implementation	Outcome	Reference
Amul, India	Reduce equipment downtime and improve efficiency	IoT-based predictive maintenance systems and automated quality monitoring.	- Reduced equipment downtime by 25%. - Maintenance costs lowered by 30%. - Milk quality consistency improved by 20%.	Amul Dairy Cooperative, 2022
Nestlé, Europe	Optimize resource usage and enhance sustainability	IoT-enabled sensors for water recycling, energy optimization, and by-product utilization.	- Water usage reduced by 20%. - Energy consumption decreased by 15%. - Whey repurposed into high-value protein supplements.	Nestlé Sustainability Reports, 2021
Voshazhnikovo Farm, Russia	Increase milk production and optimize feeding efficiency	Deployment of IoT sensors to monitor environmental conditions and cow behavior, combined with machine learning algorithms to analyze data	- 18% increase in milk production - €340,000 savings in feed costs over 180 days	Libelium, n.d.
Stellapps Technologies, India	Enhance productivity and profitability in dairy farming	Implementation of IoT-based milk production applications to monitor and manage dairy operations	- Improved productivity and profitability in dairy farming	IoT ONE, n.d.
Edge-Centric Industrial IoT Solution	Optimize dairy processing operations	Integration of IoT, Edge-AI, and smart sensors for real-time monitoring and control	- Enhanced productivity and efficiency in dairy processing	Mhapsekar et al., 2024

CHALLENGES IN IOT IMPLEMENTATION IN DAIRY PROCESSING

The dairy processing industries can benefit greatly from IoT-driven automation. Adoption and scalability may be hampered by the technical, operational, and financial difficulties associated with its deployment. The main obstacles to the use of IoT technology in the dairy sector are listed in this section.

High Initial Costs

- o Capital Investment: IoT systems involve high capital expenditures in sensors, devices, software, and network infrastructure. Most small and medium-sized enterprises (SMEs), which comprise a large part of the dairy industry, can afford these investments (Kumar & Joshi,2021). Establishing IoT systems in dairy plants increases operational costs by 15-20% in the initial years (FAO, 2023).

Data Security and Privacy

- o Cybersecurity Risks: IoT devices are prone to cyberattacks, including data breaches, ransomware, and unauthorized access. All these risks can compromise the business and product data (Mhapsekar et al., 2024).
- o Regulatory Compliance: Compliance with data protection regulations, such as GDPR and industry-specific standards, makes it more complex to implement the IoT (Libelium, n.d.).

Integration with Legacy Systems

- o Compatibility Issues: Most dairy processing facilities use aged equipment and infrastructure, and integrating modern IoT systems with them is challenging (Amul Dairy Cooperative, 2022).
- o Operational Disruptions: Integrating IoT functionality into existing systems creates short-term process disruptions (Nestlé Sustainability Reports, 2021).

Technical Competence Gaps

- o Skill Gaps: Operating and maintaining IoT systems demands advanced technical skills in sensor deployment, data analytics, and cybersecurity. Dairy operators lack this kind of skill.
- o Training Costs: Up-skilling the employees to handle IoT systems increases the operational costs and time spent (Kumar & Joshi, 2021).

Connectivity Challenges

- o Rural Locations: Dairy farms and processing units are often located in remote areas with limited internet connectivity, which hinders real-time data transmission and system efficiency (FAO, 2023).

- o Latency Issues: Poor network performance can lead to delays in data collection and analysis, reducing the effectiveness of IoT systems (IoT ONE, n.d.).

Resistance to Change

- o Cultural Barriers: Many stakeholders in the dairy industry, especially in traditional setups, resist adopting new technologies because they do not trust or understand IoT(Lee & Lee, 2015).
- o Adoption Rates: The resistance to change slows the shift to IoT-enabled systems, especially in developing regions (Gupta & Roy, 2023).

Scalability Concerns

- o Infrastructure Limitations: Expanding IoT solutions across large-scale dairy operations can strain infrastructure and resources.
- o Cost vs. Benefit: The long-term benefits of IoT investments are difficult to prove in many cases, especially for small-scale operators.

Environmental and Power Constraints

- o Energy Requirements: IoT devices consume power continuously, which may be a limitation in energy-scarce or remote locations (Mahalik & Nambiar, 2010).
- o Environmental Durability: IoT sensors and devices must endure extreme temperatures and humidity levels typical in dairy processing environments (Dabbene et al., 2014).

Addressing the Challenges

1. Cost Reduction: Incentivizes governments or organizations regarding adopting IoT for the dairy industry or offering subsidies (Kumar & Joshi, 2021).
2. Improvement of Connectivity LPWAN or Satellite-Based IoT: Low-power wide area network technologies have improved connectivity in places with little coverage.
3. Training workforce industry players can create workforce development programs that are technically skilful.
4. Cyber Security To protect data, Strong encryption procedures should be enforced, and systems should be audited regularly (Mhapsekar et al., 2024).

CONCLUSION & FUTURE DIRECTIONS

The rise in the demand for dairy products worldwide and the waste, inefficiencies, and environmental effects of conventional processing methods have put the dairy sector at a crossroads. This chapter demonstrated how IoT-driven automation in dairy processing operations may revolutionize operations by increasing output, guaranteeing resource efficiency, and cutting waste. IoT offers unparalleled footprints and can optimize processes, reduce environmental footprints, and meet consumer expectations for high-quality and sustainably produced products, from real-time milk quality monitoring to predictive equipment maintenance.

Adopting IoT has measurable advantages, as demonstrated by the case examples in this chapter, including less equipment downtime, better resource utilization, and enhanced product quality. The extensive use of IoT is not without its difficulties, though. High initial costs, technical integration issues, and data security concerns are some of the significant barriers. Addressing these issues requires collective efforts from policymakers, industry stakeholders, and researchers.

The future directions for IoT-driven automation in dairy processing are integrating advanced technologies to enhance efficiency, sustainability, and quality assurance. Automation can be leveraged with artificial intelligence (AI) and machine learning (ML) models with enhanced predictive maintenance of machinery, real-time monitoring of production parameters, and optimized supply chain logistics. Blockchain technology promises to improve traceability and transparency throughout the dairy supply chain, restoring consumer trust and regulatory compliance. In addition, introducing digital twins—the virtual replicas of physical systems—can make simulation-based testing possible for operators to identify bottlenecks and improve process efficiency without interrupting production. Further collaboration between IoT platforms and robotics in packaging and distribution can further streamline operations with reduced human intervention and consistency in product quality. Finally, these innovations will only scale if they are embedded with extreme cybersecurity measures to protect sensitive data and standardized IoT protocols in their adoption. These innovations have the promise to transform traditional dairy processing into a more connected, automated, and sustainable sector.

REFERENCES

Ahmad, T., Aadil, R. M., Ahmed, H., Rahman, U., Soares, B. C. V., Souza, S. L. Q., Pimentel, T. C., Scudino, H., Guimarães, J. T., Esmerino, E. A., Freitas, M. Q., Almada, R. B., Vendramel, S. M. R., Silva, M. C., & Cruz, A. G. (2019). Treatment and utilization of dairy industrial waste: A review. *Trends in Food Science & Technology*, *88*, 361–372. DOI: 10.1016/j.tifs.2019.04.003

Almaiah, M. A., Yelisetti, S., Arya, L., Christopher, N. K. B., Kaliappan, K., Vellaisamy, P., & Alkdour, T. (2023). A novel approach for improving the security of IoT–medical data systems using an enhanced dynamic Bayesian network. *Electronics (Basel)*, *12*(20), 1–15. DOI: 10.3390/electronics12204316

Amul Dairy Cooperative. (2022). *IoT implementation report: Efficiency and maintenance*. Internal Publication.

Arya, L., Sharma, Y. K., & Kumar, R. (2023). Towards a greener tomorrow: IoT-enabled smart environment monitoring systems. *2023 International Conference on Advances in Computation, Communication and Information Technology (ICAICCIT)*, 1112-1117. https://doi.org/DOI: 10.1109/ICAICCIT60255.2023.10465894

Arya, L., Sharma, Y. K., Kumar, R., Padmanaban, H., Devi, S., & Tyagi, L. K. (2023). Maximizing IoT security: An examination of cryptographic algorithms. *2023 International Conference on Power, Energy, Environment & Intelligent Control (PEEIC)*, 1548-1552. https://doi.org/DOI: 10.1109/PEEIC59336.2023.10451210

Aung, M. M., & Chang, Y. S. (2014). Traceability in a food supply chain: Safety and quality perspectives. *Food Control*, *39*, 172–184. DOI: 10.1016/j.foodcont.2013.11.007

Dabbene, F., Gay, P., & Tortia, C. (2014). Traceability issues in food supply chains: A review. *Biosystems Engineering*, *120*, 65–80. DOI: 10.1016/j.biosystemseng.2013.09.006

Fitcows. (n.d.). Milk monitoring system using IoT to check milk's quality. Retrieved from https://www.fitcows.com/blogs/milk-monitoring-system-using-iot/

Food and Agriculture Organization (FAO). (2023). Global milk production statistics. Retrieved from https://www.fao.org/dairy-production-products/production/milk-production/en

Galukande, E., Mulindwa, H., Wurzinger, M., Roschinsky, R., Mwai, A. O., & Sölkner, J. (2013). Cross-breeding cattle for milk production in the tropics: Achievements, challenges, and opportunities. *Animal Genetic Resources*, *52*, 111–125. DOI: 10.1017/S2078633612000471

Gupta, P., & Roy, S. (2023). Enhancing dairy productivity with IoT. *Advances in IoT Systems*, *5*(4), 299–315.

IoT ONE. (n.d.). IoT-based milk production application. Retrieved from https://www.iotone.com/case-study/iot-based-milk-production-application/c602

Kumar, A., & Joshi, R. (2021). IoT in dairy: A sustainable approach. *International Journal of Dairy Technology*, *74*(3), 487–495.

Lee, I., & Lee, K. (2015). The Internet of Things (IoT): Applications, investments, and challenges for enterprises. *Business Horizons*, *58*(4), 431–440. DOI: 10.1016/j.bushor.2015.03.008

Libelium. (n.d.). How a dairy farm increased their milk production 18% with IoT and machine learning. Retrieved from https://www.libelium.com/libeliumworld/success-stories/how-a-dairy-farm-increased-their-milk-production-18-with-iot-and-machine-learning/

Mahalik, N. P., & Nambiar, A. N. (2010). Trends in food packaging and manufacturing systems and technology. *Trends in Food Science & Technology*, *21*(3), 117–128. DOI: 10.1016/j.tifs.2009.12.006

Mhapsekar, R. U., O'Shea, N., Davy, S., Kilbane, D., & Abraham, L. (2024). An edge-centric industrial IoT solution for smart dairy processing. *IEEE Internet of Things Magazine*, *7*(5), 80–87. DOI: 10.1109/IOTM.001.2400010

Nagaraj, S., Kathole, A. B., Arya, L., Tyagi, N., Goyal, S. B., Rajawat, A. S., Raboaca, M. S., Mihaltan, T. C., Verma, C., & Suciu, G. (2022). Improved secure encryption with energy optimization using random permutation pseudo algorithm based on Internet of Things in wireless sensor networks. *Energies*, *16*(8), 1–16. DOI: 10.3390/en16010008

Nestlé Sustainability Reports. (2021). Water and energy optimization in dairy processing facilities. Retrieved from https://www.nestle.com

Shamsuddoha, M., & Nasir, T. (2024). Utilizing dairy waste processing for organic agricultural production: A sustainable approach to producing organic goods. *Processes (Basel, Switzerland)*, *12*(11), 2521. DOI: 10.3390/pr12112521

Smith, L., & Wang, Z. (2020). Waste reduction in dairy processing through IoT automation. *Journal of Sustainable Agriculture*, *12*(2), 200–214.

Zarin, I., Talukder, A. K., Hoque, M. N., Rahman, M., Haider, G., Rahman, A. N. M. A., & Das, Z. C. (2023). Current scenario on reproductive management and performances of dairy cows in Manikganj district. *Annals of Bangladesh Agriculture*, *27*(2), 129–144. DOI: 10.3329/aba.v27i2.72541

KEY TERMS AND DEFINITIONS

Cold Chain Management: Temperature-controlled logistics enabled by IoT to preserve the quality and shelf life of dairy products during distribution.

Digital Twin: A virtual model of a physical system used to simulate, predict, and optimize operations in dairy processing without interrupting actual workflows.

Edge Computing: Data is processed at or near the source (e.g., sensors or machines) rather than in centralized data centers, ensuring low latency in decision-making.

Energy Efficiency: The practice of using less energy to perform the same tasks during dairy processing, often achieved through IoT-enabled automation, real-time monitoring, and optimization, resulting in reduced operational costs and lower environmental impact.

Homogenization: A dairy process that breaks down fat molecules in milk to create a uniform texture, often controlled by IoT pressure sensors.

Internet of Things (IoT): A network of linked devices with sensors, software, and connections built to gather and share data in real time for automation and decision-making is known as the Internet of Things (IoT).

Microbial Detection System: IoT-based sensors that detect and quantify microbial content in milk or dairy products to ensure hygiene and safety.

Pasteurization: Pasteurization is the process of heating milk to eradicate bacteria. IoT enables precise temperature control for effective pasteurization.

pH Sensor: An IoT-enabled sensor that monitors the acidity of milk or dairy products, ensuring freshness and product stability.

Quality Assurance: A set of procedures enabled by IoT to maintain consistent product standards by monitoring and controlling quality parameters in real-time.

RFID (Radio-Frequency Identification): Technology used to identify and track tags attached to dairy containers, enabling end-to-end supply chain monitoring.

Resource Optimization: Resource optimization is the strategic application of automation and analytics to reduce raw material, energy, and water usage during the dairy processing.

Smart Farming: Using IoT technologies in livestock and milk production to improve productivity, animal health, and operational efficiency is known as "smart farming."

Supply Chain Traceability: The ability to track a dairy product's journey across the supply chain, from farm to consumer, using technologies like RFID and blockchain for transparency and accountability.

Temperature Sensor: The device continuously monitors milk temperature during collection, storage, and transportation to prevent spoilage.

Chapter 7
Integration of Blockchain in the Dairy Sector Value Chain

Jeremy Issa Victor
https://orcid.org/0009-0000-3959-1125
CML Global Consulting, Morocco

ABSTRACT

This chapter examines the transformative potential of blockchain technology in the dairy sector, with a focus on camel milk as an emerging market. It explores how blockchain enhances traceability, quality assurance, and transparency while addressing challenges like inefficiencies, food safety, and inequities faced by small-scale producers. Tools such as IoT integration and smart contracts are highlighted for their role in streamlining operations and ensuring compliance with safety and sustainability standards. The chapter also discusses blockchain's ability to reduce intermediaries, promote fair trade, build consumer trust, and expand market access for niche products like camel milk. Challenges to adoption, including technological barriers, costs, and regulatory concerns, are explored alongside future prospects, such as innovations in scalability, emerging market expansion, and collaborative efforts between governments and industries to create a more equitable and sustainable dairy value chain.

DOI: 10.4018/979-8-3693-9949-1.ch007

INTRODUCTION

Overview of the Dairy Sector's Current State

Dairy's a big deal—most of us don't think about it, but it really is. We're talking about a system that feeds a huge part of the world. People rely on milk, yogurt, cheese, butter—every single day. It's part of daily life, and it supports a lot more than just nutrition. Jobs, trade, local and national economies—it's all tied in. The numbers? Massive. The industry pulls in over $700 billion a year. And behind that are millions of people: farmers, drivers, factory workers, retailers. The whole chain.

But here's the thing—just because it's big doesn't mean it's running smoothly. In fact, that size brings problems.

For one, the way dairy products move around is incredibly complex. Milk usually comes from all over—different farms, often in remote areas. Then it has to get to a facility, be processed, packed, and sent out again. It can go halfway across the world before someone finally drinks it. That kind of setup? It leaves a lot of room for mistakes. Products can spoil. Deliveries can get messed up. And if there's ever a food safety issue, tracing where the problem started can take ages. People are also more skeptical now—about what's in their food, where it came from, and whether labels are even honest.

And that's just logistics. There's a bigger conversation around how dairy impacts the planet. Industrial farming uses huge amounts of water. It contributes to greenhouse gas emissions. And over time, the land suffers. Plus, while some farms are thriving, a lot of smaller ones aren't. Especially in places with fewer resources, small farmers often can't compete. They don't get good market access and usually depend on middlemen who don't always play fair.

So yeah, it's a tough situation. But there's a tech angle here that's starting to get real attention—blockchain.

Why Does Blockchain Matter in Global Value Chains?

I know, it sounds like crypto jargon. But stripped down, it's just a digital record system that can't be easily altered. And the cool part? Everyone in a supply chain can access the same data. No one can go back and fudge the numbers later. That's a big deal for transparency.

In dairy, this could mean a few things:

You'd be able to track a product from the cow to the shelf. Literally. That helps during safety scares—if there's bad milk, you know exactly where it came from.

> You could hook up sensors that log storage temperature or delivery times and connect them to the blockchain. So if something spoils, it's easier to see why.

For consumers, it's all about trust. People want to know their food is legit, that it came from somewhere ethical. Blockchain makes that info available—quickly and clearly.

And maybe most importantly, this tech can help farmers. Without so many layers between them and the buyer, they can earn more and get paid faster. That alone could change things for a lot of small producers.

Sure, it's not going to fix everything overnight. But if we're serious about building a better, more honest food system, tools like this might be part of the answer

Camel Milk as a Growing Industry: Unique Opportunities and Challenges

Camel milk is stepping into the spotlight. Long a staple in desert communities, it's now gaining global traction as more people discover its nutritional and environmental benefits. And while blockchain and other tech tools offer a lot to the dairy industry as a whole, camel milk presents a unique case—one full of opportunity, but also challenges.

So what's driving this shift?

For starters, the nutritional profile of camel milk is hard to ignore. It's rich in vitamins and minerals, packed with bioactive compounds, and naturally lower in lactose—making it easier to digest for people with lactose intolerance. On top of that, it has antimicrobial properties that have drawn the interest of health-focused consumers worldwide.

Opportunities in Camel Milk

We're seeing growing demand in regions far from traditional camel-rearing areas—places like Europe, North America, and Asia. Consumers in these markets increasingly view camel milk as a premium health product. It fits neatly into the lifestyle of people who want clean, functional foods that offer more than just basic nutrition.

There's also a clear sustainability advantage. Camels thrive where other livestock struggle. They need less water, less feed, and they're well-adapted to harsh climates. In an era where droughts, heatwaves, and resource scarcity are becoming the norm, that matters. Camel milk production has the potential to be a more eco-responsible alternative to conventional dairy, especially in dry and semi-arid regions.

And while camel milk still represents just a sliver of the global dairy market, its niche appeal is exactly what makes it valuable. It's not trying to compete with cow's milk at scale—it's carving out its own space as a high-value, specialty product. For producers and entrepreneurs, that's where the growth potential lies: in creating quality over quantity, and building brands around authenticity, wellness, and sustainability.

That said, there's still work to do. Infrastructure is limited in many producing regions. International standards and regulations are still catching up. And cold chain logistics—getting fresh camel milk to distant markets—remains a hurdle. But none of that cancels out the fact that camel milk is on the rise, and the industry around it is maturing fast.

For anyone watching the future of food, this is one space to keep an eye on.

Challenges in Camel Milk Production

Working with camel milk sounds promising—and in many ways, it is. But the truth is, producing it consistently and at scale isn't as easy as it might look from the outside. There are a few big hurdles that anyone in the space knows all too well.

To begin with, camels just don't produce that much milk. Not compared to cows, anyway. Their lactation periods are shorter, and the daily output is lower. So even if the demand is growing, supply can't just be scaled up overnight. That's a tough starting point.

And then you've got the reality that most camel milk still comes from very remote places—often from nomadic or semi-nomadic herders. These producers might be days away from a processing center, and the infrastructure to collect and transport milk quickly (and hygienically) just isn't always there. The distances, the heat, the logistics—it all adds up.

Also, let's be honest: processing infrastructure is spotty at best. In many regions, there's no consistent standard for how camel milk gets stored, pasteurized, or packaged. Without proper refrigeration or a working cold chain, a lot of good milk goes to waste. And what does get to market? It can vary a lot in quality, which doesn't help when you're trying to build a new product category.

And here's another thing: most people have never tried camel milk. They might've heard of it—or seen something online—but they don't really know what it is, why it matters, or what makes it different from cow's milk. So, even if it reaches store shelves, there's still this big awareness gap to overcome.

The Role of Blockchain in Camel Milk

So what could actually help?

Well, blockchain might sound like a buzzword—but in this case, it's got some real-world potential. Not because it's trendy, but because the problems it solves are the exact ones the camel milk world is facing.

First, traceability. Imagine being able to track a single bottle of camel milk all the way from the herder's hand to the grocery store shelf. That kind of visibility could build trust with customers—especially in countries where camel milk is still a curiosity.

And if you plug in real-time data tools—like temperature sensors during transport—and feed that into the blockchain? Now you've got a record of exactly how that milk was handled. No more guessing about freshness or safety. That's a game-changer for quality assurance.

It could also help level the playing field for the producers. Right now, small herders often depend on middlemen to get their product to market, and that cuts into their earnings. With blockchain, you can cut through that chain—letting herders connect directly to buyers. Fairer prices, faster payments, fewer unknowns.

And there's another upside too: educating the consumer. A simple QR code on the bottle could tell the whole story—where the milk came from, who produced it, what the nutritional benefits are. That's not just marketing fluff—it's a way to make the product real to people who've never seen a camel in their life.

So sure, camel milk has its share of obstacles. But the right tech—used smartly—could turn a lot of those obstacles into stepping stones. It's not about replacing tradition. It's about helping it scale, survive, and thrive in a new kind of market.

BLOCKCHAIN TECHNOLOGY: A BRIEF OVERVIEW

Definition and Core Principles of Blockchain

Okay, so let's try to make sense of this whole "blockchain" thing—because it's one of those words that gets thrown around a lot, but still confuses people. At its core, it's actually pretty simple: it's a kind of record-keeping system, but instead of everything being stored in one place, the info is shared across a bunch of computers. No central boss or authority, no single version of the truth sitting on one server.

This idea—decentralization—is a big part of what makes blockchain so powerful. Everyone in the system has a copy of what's going on, so if one computer crashes or someone tries to cheat, the others keep things honest.

What's really interesting is how the data gets locked in. Once you add something to the blockchain, it's there. It doesn't get edited or quietly changed later. That's why people say it's "immutable." It's not impossible to change, technically, but you'd need a ton of people to agree, and that just doesn't happen in practice.

And here's the thing—because everyone sees the same thing, there's a sense of shared trust. Even if people don't know each other, they can rely on the data being legit. That's what makes it work in supply chains, especially when there are lots of hands touching a product along the way.

Now, security-wise, blockchain's no joke. It uses pretty intense encryption, and the way things get added—there's this process where the network has to agree it's valid first. That can look different depending on the system (some use something called Proof of Work, others use Proof of Stake), but the goal is the same: make sure no bad data gets through.

Oh—and then there are smart contracts. These are basically little programs baked into the system. They say, "If this happens, then do that." And they just... do it. No back-and-forth emails, no middlemen. It's all automated. That part has a ton of potential in places like agriculture or food trade, where steps need to happen in a certain order and timing matters.

Anyway, blockchain's not magic, but it does offer something we've been missing in a lot of industries: a shared, trustworthy way to track what's happening—without needing to rely on one single company or authority to tell everyone what's true.

Relevance of Blockchain to Supply Chain Management

Supply chains sound neat on paper, but anyone who's worked with them knows how messy they actually are. Dozens—sometimes hundreds—of people involved, across countries and systems that barely talk to each other. Things get lost. Data doesn't line up. One mistake at the start and suddenly there's a bottleneck three stages down the line. That's just normal in this world.

Blockchain has been getting attention not just because it's "new," but because it actually solves some of this chaos in a pretty straightforward way: it keeps records that can't be faked and lets everyone involved see the same thing at the same time.

Enhancing Traceability

One of the biggest headaches in supply chains is traceability. You want to know where something came from, who handled it, and whether it was stored properly. Usually, that info is buried in someone's spreadsheet or handwritten in a notebook in another country. But with blockchain, you can log each step—from the raw material to the final sale—and that record doesn't get altered. So if there's a problem, you can figure out what happened and when, without the usual runaround.

Let's take camel milk as an example. It's perishable. It moves through a lot of hands. And most of it comes from remote areas. If you could scan a bottle and see which herder it came from, what date it was collected, how it was stored, and how long it was in transit—well, that's a big deal. That's not just interesting info. It's what builds trust with buyers who don't know the source firsthand.

Improving Transparency

There's also the transparency angle. And it's not just about honesty—it's about coordination. When everyone in a supply chain is working off the same set of facts, a lot of confusion goes away. And when that data is recorded automatically and can't be tweaked, it's a lot harder to fudge numbers or make things look better than they are. For a product like camel milk, where ethical sourcing and origin matter, being able to prove where it came from and how it was handled is powerful.

Strengthening Food Safety

Another issue, especially in food, is safety. Milk spoils. Things can go wrong in storage. With blockchain hooked up to sensors—like temperature trackers—you can catch a problem while the product's still moving. That means instead of finding out too late, producers or transporters can act fast, avoid losses, and protect the brand.

Streamlining Logistics and Reducing Inefficiencies. There's also something to be said about cutting out the middlemen. A lot of delays and costs in supply chains come from people and paperwork between point A and B. But blockchain, especially when paired with smart contracts, lets people interact directly. You could have camel herders selling directly to buyers, getting paid automatically once their delivery is confirmed. That's a huge shift—not just in efficiency, but in fairness.

Building Consumer Trust

And if we zoom out a bit, there's a sustainability piece too. When you can track exactly how something was produced and moved, you can also measure its footprint. In camel milk, that might mean knowing how much water went into feeding the animals, how far the milk traveled, or what emissions were involved in storage and transport. That's info that companies can use to improve—and consumers can use to make more informed choices.

Enabling Sustainability

At the end of the day, blockchain doesn't fix every problem in the supply chain world. But it gives us tools we've never had before—tools that make things more open, more efficient, and a lot more fair. And for industries like dairy, especially ones just starting to grow internationally, that could change everything.

CHALLENGES IN THE DAIRY SECTOR

The dairy industry plays a huge role in feeding people across the globe, but it's also under strain. It's complex, often outdated in how it operates, and riddled with problems that touch every part of the chain—from the farm all the way to the store shelf. And honestly, if we don't find better ways to manage things like traceability, safety, and fairness, it's only going to get harder to keep up with rising demand and consumer expectations.

Traceability Issues

Let's start with traceability. Everyone says they want to know where their food comes from, and for good reason. But the truth is, it's often hard to answer that question—especially in dairy. Milk usually gets collected from a bunch of different farms, all pooled together before it's even processed. Once it's mixed, it's basically impossible to tell exactly where it came from. So if there's a contamination issue or a question about quality, good luck tracing it back. Add international trade into the mix, and the problem only grows. Dairy moves through so many hands—suppliers, shippers, processors, distributors—that there are plenty of places where data can get lost or, worse, changed. Sometimes that's by mistake. Sometimes it's on purpose.

And then there's the consumer side. People are asking more questions now. They want to know: Is this really organic? Was it ethically sourced? Did the farmer get paid fairly? When brands can't answer those questions clearly, they lose trust—and once you lose that, it's hard to win it back. Camel milk producers, in particular, face this challenge on a whole other level. The way camel milk is collected—usually in remote, scattered areas by independent herders—makes documenting the milk's journey incredibly tough. There's just not much infrastructure in place to track anything in real time.

Food Safety Concerns

Now, food safety. Dairy is fragile. It spoils easily. A small mistake in handling or a temperature drop during transport, and the whole batch might be useless—or worse, unsafe. Bacteria like E. coli or Salmonella don't need much of an opening to cause problems, and if you don't catch contamination early, it can slip through the entire chain unnoticed. A lot of cold chains still aren't reliable, especially in developing regions. The refrigeration breaks down, the transport takes too long, or the storage isn't up to standard. And because traditional systems aren't built for real-time monitoring, problems often don't get noticed until it's too late. At that point, the product's already on store shelves—or worse, in someone's home.

Camel milk has even more hurdles in this area. It needs specialized processing, and many regions just don't have the right facilities or cooling infrastructure. Even if the milk is high quality at the source, by the time it reaches the buyer, that quality can degrade quickly without the right systems in place.

Inefficiencies in Supply Chains

On top of all that, supply chains themselves are often inefficient and way too complicated. Milk doesn't just go from farm to fridge. It might pass through five or six layers of middlemen before reaching the consumer. Every layer adds time, cost, and the chance for something to go wrong. And a lot of operations are still using paper records or basic manual systems that are slow and full of errors. It's 2025, but in some places, the processes haven't changed in decades.

Waste is another huge issue. Milk goes bad, shipments are delayed, and a lot of product never makes it to market. According to the FAO, a sizable chunk of the world's milk is lost before it's even sold—and that's not just money down the drain, it's food, time, and resources wasted. In the camel milk sector, these inefficiencies are even worse because the supply is so limited to begin with. Production levels are low, and the distances between producers and buyers are often long. The market is still niche, which makes every logistical decision a little more expensive, a little more delicate.

And then there's the elephant in the room: small producers. They're the backbone of the dairy world, especially in places like Africa, the Middle East, and parts of Asia. But they're also the ones facing the steepest climb. Most don't have access to capital. Many don't have reliable tech or basic infrastructure. They're cut off from big markets and forced to sell to middlemen who offer the lowest prices and pocket the rest. Camel herders feel this even more—isolated, under-supported, and without collective bargaining power. Many work alone, without the benefits of cooperatives or producer associations that could help them get better deals or share resources.

Training is another gap. A lot of small-scale farmers don't get the information they need to keep their animals healthy or their milk clean. They're doing their best, often relying on inherited knowledge, but without up-to-date tools or support, quality suffers. And that means they stay stuck—excluded from high-value markets and vulnerable to every price drop or market shift that comes along.

All of this paints a picture of an industry that's working hard—but still falling short. From the way data is tracked, to how products are handled, to who actually gets paid fairly—there are cracks everywhere. And patching them one by one isn't enough.

That's why new approaches, especially ones using tech like blockchain, are starting to get serious attention. Because if there's a way to make the system more transparent, more fair, and better connected, then maybe there's also a way to finally give small producers a seat at the table—and make the entire dairy sector stronger and more sustainable for everyone involved.

THE POTENTIAL OF BLOCKCHAIN IN DAIRY

The dairy industry has been trying to modernize for years, but many of the same challenges keep surfacing—lack of transparency, messy supply chains, and unfair systems that push small producers to the margins. Blockchain might not be a silver bullet, but in this context, it actually feels like a tool with real weight. Not because it's trendy, but because it fits the problems.

At its core, blockchain creates a shared system of records that everyone involved can trust. That's powerful—especially when you're talking about something as complex and fragile as dairy. Every time milk changes hands—from the farm to the collector, to the processor, to the shelf—that event can be logged. Not on paper. Not in an Excel file tucked away in someone's office. But in a permanent, verifiable system that can't be edited quietly afterward.

Enhanced Traceability and Transparency

Take camel milk, for example. It's usually collected in rural, often isolated regions, where record-keeping is inconsistent or nonexistent. With blockchain, you can trace each batch from the exact herder who milked it, through every stage of its journey. If there's ever a safety concern or a need to verify sourcing, that information is already there. No need for long investigations or guessing games. The trace is built in.

And for consumers, this isn't just a backend upgrade—it changes the experience. Imagine picking up a bottle of camel milk at a store and scanning a QR code to see not only where it came from, but how it was stored, how long it took to arrive, and even whether the camels were raised with good animal welfare standards. That's not futuristic—that's already happening in some pilot programs. And when people can see that kind of information, they're more likely to trust what they're buying—and the people behind it.

It also makes a real difference when things go wrong. Contamination, for instance. In traditional systems, when something is off, the best you can do is guess and recall a wide range of products just to be safe. But with blockchain? You can pinpoint the exact batch that had an issue, pull only what's necessary, and respond quickly. It saves time, money, and most importantly, protects public health.

Quality Assurance and Compliance

On the quality side, things get even more interesting. If you add smart sensors into the process—tracking temperature, humidity, pH, and so on—you can start feeding live data into the blockchain too. So if camel milk gets too warm during transport, that gets logged. It's visible. And it can trigger actions automatically. For instance, a smart contract could be set up to stop a shipment from being released for sale if it doesn't meet certain conditions. No more crossing your fingers that the cold chain held up.

That kind of system doesn't just prevent problems—it also rewards good practices. If a producer consistently meets quality thresholds or follows sustainability guidelines, they can prove it. And that opens the door to better pricing. Buyers can see the data and pay accordingly. It becomes easier to separate the high-quality suppliers from the rest, and that encourages ongoing improvement.

Reducing Intermediaries and Ensuring Fair Trade

Another major upside is the way blockchain simplifies the entire payment and logistics system. Right now, dairy supply chains are clogged with intermediaries—agents, brokers, handlers, you name it. Each adds a layer of cost, and often confusion. With blockchain, you can shrink the chain. Producers can interact more directly with buyers, get paid faster, and keep more of the value they generate.

And that's not just theory. In a blockchain-based system, a camel herder in a remote area could deliver milk, have the weight and quality logged on-site, and trigger a payment the same day. No more relying on verbal agreements or waiting weeks for settlement. Everything is recorded, everything is transparent.

This matters especially in markets like camel milk, where most producers are small-scale and disconnected from the larger economy. Giving them access to fair systems—and to data they can actually use—starts to shift the balance a bit. It creates space for fair trade to become real, not just a label.

And finally, there's the bigger picture. Companies and consumers alike are starting to care about sustainability in a serious way. Blockchain allows for tracking that too—things like water use, feed sourcing, carbon emissions. If a buyer wants to support a low-impact product, or a producer wants to showcase how they're reducing waste, there's a way to do that with actual proof behind it.

So no, blockchain won't solve everything overnight. But what it can do is bring a level of visibility, accountability, and fairness that the dairy world—especially its more vulnerable corners—has been missing for a long time. If we're serious about building a more equitable and resilient food system, this is a tool worth taking seriously.

MILKING TRUST: HOW BLOCKCHAIN CAN UNLOCK DAIRY'S GLOBAL FUTURE

Camel milk has been a quiet constant in the diets of desert communities for generations. It's not a trend there — it's part of life. But lately, far beyond the dunes, interest has been picking up in places where most people have never even seen a camel. In cities from Toronto to Berlin, health food stores are starting to stock it. There's buzz around its nutritional profile, its digestibility, and even its supposed benefits for people with diabetes or autism. Suddenly, what was once local and traditional is turning into something premium — a dairy product with global aspirations.

The Exponential Potential of Camel Milk in Global Markets

And it's not just hype. Nutritionally, camel milk really does stand out. It's got more vitamin C than cow's milk, more iron, and a different set of fats — not to mention those bioactive compounds that researchers are still learning about. It's easier to digest for many people, especially those with lactose intolerance or cow milk protein allergies. All of that makes it a good fit for today's growing market for functional, wellness-driven foods.

But if you zoom out a bit, you start to see the cracks in the system that's supposed to bring this milk from remote desert outposts to modern global shelves. The product is valuable. The market is real. The demand is growing. And yet, camel milk still accounts for barely a fraction of a percent of global dairy trade. That's not because people don't want it — it's because getting it there is hard.

Specific Challenges in Camel Milk Production

One of the first hurdles is yield. A single camel just doesn't produce much milk compared to a cow. Maybe a few liters per day, sometimes less, and only during a certain part of the year. So supply is tight from the start. But it's not just about volume. Most camel milk is produced in some of the hardest-to-reach areas on Earth, by herders who live nomadic or semi-nomadic lifestyles. They move with their animals. They don't always have cold storage. And they often don't have reliable internet or access to the kinds of supply chain systems most companies take for granted.

You add that all up — low output, remote producers, fragile infrastructure — and you start to see why camel milk hasn't "scaled" the way other niche health products have. It's not that it can't. It's that the usual tools don't fit.

Blockchain as a Foundational Solution for Camel Milk Supply Chain Development

Blockchain, in that context, isn't just another digital upgrade. It's something that actually makes sense for the shape of the problem. It offers a way to build trust across a supply chain that's often invisible. It provides a framework where everyone — even someone far out in the desert with limited access — can still be part of the same system.

Let's say milk is collected in a rural village in Mauritania. Right now, there's a good chance that the herder has no idea where that milk will end up — who will process it, sell it, drink it. And on the flip side, the person buying a bottle of camel milk in a shop in London has no clue where it came from, how it was handled, or whether it was kept at the right temperature along the way.

End-to-End Traceability

With blockchain, that changes. Every time that milk moves — from the herder to the collector, from the truck to the cold storage, to the packaging line — that movement is logged. Not by hand, not in some notebook that could go missing, but in a digital record that can't be edited later. Add sensors and real-time data feeds to the equation, and you start to get a clear, verifiable picture of quality, temperature, timing, and storage conditions.

If something goes wrong — let's say a shipment wasn't chilled properly — the system flags it. And if things go right, that record becomes a kind of proof-of-care. A small producer can say, "Look, here's what I delivered, when I delivered it, and how it was handled." They don't need to rely on vague agreements or middlemen who take too much. That kind of visibility helps even the playing field, especially for small-scale herders who've historically had little control over how their product is valued.

Quality Assurance

For consumers, too, this isn't just technical. It's personal. Food is identity. And more and more, people want to know where their food comes from — not just the location, but the story. With blockchain, you can give them that. A QR code on a bottle might lead them to the name of the herder, the region, the animal welfare practices used, even sustainability stats — like how much water was used or whether plastic packaging was avoided.

That kind of storytelling doesn't just build loyalty. It builds value.

Empowering Small Producers

It also helps with something that's often overlooked: market positioning. If camel milk producers want to compete in premium markets, especially in Europe or North America, they'll eventually need certifications — organic, fair trade, sustainable, and so on. With blockchain-backed records, that certification process becomes easier. It's not based on trust alone. It's based on data.

Even from an operations standpoint, the benefits are hard to ignore. Automating compliance, tracking transport, minimizing spoilage — it all leads to less waste and higher profit margins. And in a business where supply is inherently limited, reducing waste can matter just as much as increasing output.

There's still work to do, of course. Blockchain systems need to be adapted to local realities. They need to be accessible, user-friendly, and fair. You can't just parachute in technology and expect it to stick. But the blueprint is there. And for camel milk — a product rich in history but short on infrastructure — blockchain might be one of the few technologies that's not only practical, but deeply compatible with the culture around it.

The more you look at it, the more it becomes clear: camel milk doesn't need a revolution. It needs a better bridge. Blockchain, if implemented thoughtfully, can be that bridge — connecting producers with markets, stories with trust, and tradition with global opportunity.

BLOCKCHAIN IMPLEMENTATION: PROCESSES AND TOOLS

Implementing blockchain in the dairy world isn't just a matter of flipping a switch. It's not a ready-made plug-in. In fact, getting it right means working with the chaos—not against it. That's especially true in places like camel milk production, where operations aren't centralized, and the supply chain starts miles away from anything that looks like a networked factory or refrigerated truck.

Still, if there's one thing that makes blockchain so valuable here, it's how flexible it can be—when paired with the right tools and approached the right way. This isn't just about digital ledgers. It's about tying technology into physical, everyday systems in a way that makes life easier for the people doing the hard part: collecting, storing, transporting, and trying to sell perishable milk in a demanding market.

Integrating IoT (Internet of Things) Devices

Let's start with the physical side—IoT devices. These aren't fancy concepts anymore. They're small sensors, thermometers, GPS units, things you can attach to a truck or a storage container or even a milk can. The idea is simple: track the things that matter in real time. In dairy, that usually means temperature and hygiene. In camel milk? Temperature is everything.

You'd be surprised how quickly camel milk can spoil. The nutritional profile that makes it so appealing—bioactive compounds, unprocessed freshness—also makes it more fragile. A few hours without cooling can ruin a batch. So if a sensor logs that a shipment sat in 32°C heat for too long, you need to know. Not tomorrow. Now. And with blockchain, that info doesn't just get logged. It gets locked. It can't be changed or erased, and that kind of permanence matters when people start pointing fingers after something goes wrong.

But beyond just alerts, these devices create a trail. They generate data—location, temperature, handling—and that data gets uploaded automatically, creating a digital footprint for every single container. From the moment milk is collected until it hits a shelf, you've got a living record. And everyone involved—herder, processor, distributor—can access it. That's traceability, but it's also protection. It protects the buyer, yes, but it also protects the seller. Especially when the seller is someone who normally has no voice in the system.

Smart Contracts for Quality Assurance

And then there are smart contracts. This is where it gets really interesting.

A smart contract is basically code that sits inside the blockchain and says, "If this happens, then do that." If the milk stays under 8°C the whole way, release the payment. If not? Flag it. Block it. Or send it for extra testing. That kind of automation saves time, but more importantly, it cuts out a lot of grey areas. Decisions get made on facts, not assumptions.

And this has real effects, especially on pricing. Imagine a camel herder whose milk always arrives cold, clean, and on time. Right now, that quality might go unnoticed—or worse, be diluted in a pool of other milk that didn't meet the same standard. But with blockchain and smart contracts, their individual contribution is visible. It's recorded. And it can be rewarded—immediately, with no need to chase down payments or argue with buyers about who did what.

That kind of fairness? It's rare in traditional supply chains. But it becomes possible when the system doesn't rely on trust alone—it relies on verified data.

Steps to Adopt Blockchain Across Value Chains

But again, getting there doesn't happen overnight. Rolling out blockchain in a dairy supply chain—especially one as scattered and unpredictable as camel milk—needs a process. And it needs buy-in.

Identifying Pain Points. The first step? Start with the problems. That sounds obvious, but it's easy to skip. Where is the supply chain breaking down? Is it in transport? Storage? Is payment too slow? Is quality control inconsistent? In camel milk, the pain points usually start with logistics. The milk is produced far from cities, far from factories, and often moves long distances before it's even cooled.

Building a Collaborative Ecosystem. Once the weak spots are mapped, the next move is to gather the right people. No blockchain system works if it's imposed top-down. Everyone from herders to regulators to retailers needs to be part of the build. Not just consulted. Engaged. That includes technology partners too—folks who can tailor blockchain platforms and IoT tools to the context, not just dump software and leave.

Choosing the Right Blockchain Platform. Picking the blockchain platform itself isn't trivial either. Public systems like Ethereum are open, sure, but they might not be right for something that involves sensitive supply and pricing data. In cases like dairy, especially camel milk, a permissioned blockchain usually makes more sense. That way, access is granted to those who need it, and sensitive information doesn't leak into the wild.

Integrating IoT and Smart Contracts. Next comes the gear—the actual deployment of sensors and contracts. IoT devices have to go into trucks, coolers, farms. Smart contracts need to be designed around real, testable events. Not "Did the milk arrive in good shape?" but "Was it stored below X temperature for the entire trip?" Specific, measurable, enforceable.

Piloting the System. Then it's time for a pilot. Start small. Maybe just one route. One processor. One cooperative of camel herders. Run the system. Watch what breaks. Watch what works. Document everything. Feedback here is gold—it's where digital meets dirt, where code meets camel.

Training Stakeholders. And don't skip the training. Even the most elegant system will fail if users don't understand it—or fear it. Training has to be contextual. It has to consider literacy, language, tech familiarity. You can't roll out blockchain tools that only work for tech-savvy users when your producers are rural herders with no prior exposure to digital systems. The learning curve has to be built in. And it needs to be respectful of how people already work.

Scaling the Solution. Once the pilot holds, then—and only then—can you scale. Slowly. Carefully. Add regions. Add features. Integrate retailers. Loop in certifiers. At this stage, continuous feedback and monitoring are just as important as they were in the beginning.

Monitoring and Optimizing. And yes, none of this is cheap. Or easy. But the value isn't just in making the supply chain faster. It's in making it more trustworthy, more balanced, and more supportive of the people at the base of the pyramid. It's about giving camel milk a digital spine strong enough to carry it into global markets.

The right tools—real-time sensors, transparent ledgers, automated enforcement—don't just reduce waste and streamline operations. They create a system where the herder with a few camels and a tin bucket isn't excluded, but connected. Where what they do, and how they do it, finally counts.

ECONOMIC AND SOCIAL IMPACTS

When you start talking about blockchain in the dairy sector, it's easy to get stuck in the technical weeds—ledgers, smart contracts, sensors, data layers. But the real story—the one that matters most—is about people. Behind every carton of milk, there's a herder, a farmer, a driver, someone whose livelihood depends on a system they rarely have power in. And that's where blockchain starts to change things—not just as a technology, but as a tool for fairness, inclusion, and resilience.

Impact on Small-Scale Producers

In places where dairy production is small-scale and scattered—like with camel milk—producers often find themselves shut out of real value. They do the hardest work, but see the smallest return. They're the last to get paid, the first to get blamed when something goes wrong, and they have little say in how their product is priced, moved, or even labeled. The system has been skewed against them for decades.

What blockchain offers is a way to shift that balance. By making supply chains more visible and traceable, it levels the playing field. A camel herder who used to sell milk blindly into a system now has tools to show exactly what they delivered—when, how much, and at what quality. That data can trigger payment automatically, based on transparent standards, not vague promises. It also means they can prove they met hygiene or cooling requirements, and be rewarded for it. No more depending on middlemen to tell their story. The proof is built in.

And with that visibility comes access. Many small-scale producers never reach premium markets—not because their product isn't good, but because the systems they're in aren't trusted. Exporters want traceability. Certifications. Records. Blockchain makes it possible to meet those standards. Suddenly, a product like camel milk, often produced in remote, resource-poor areas, can reach buyers in Europe or Asia looking for something unique, ethical, and clean.

But it's not just about selling more. It's about knowing more. With the right systems in place, producers get access to data—real-time info on prices, demand trends, storage performance, even buyer preferences. That feedback loop helps them improve. They can make better decisions, adjust feeding practices, manage herds more efficiently, and ultimately produce better milk. That kind of insight has been reserved for big industrial players. Blockchain brings it down to earth.

It also stops exploitation. When every transaction is logged, and pricing is visible, it becomes a lot harder for bad actors to skim profits or lie about terms. Everyone sees the same numbers. Everyone knows what was agreed to. That's not just fairer—it's safer. Especially in industries like camel milk, where producers have historically worked in isolation and vulnerability.

Building Consumer Trust

On the other side of the chain, consumers benefit too—and not just in an abstract way. People today want more than just "milk." They want to know where it came from, how it was handled, whether the animals were treated well, whether the workers were. And blockchain makes that visible. A simple scan on a label can tell the whole story—from desert to fridge.

That's powerful. Not just for marketing, but for building trust. When someone chooses camel milk, especially in a premium health-food store thousands of miles from the nearest camel, they're buying more than nutrition—they're buying peace of mind. Knowing that the product is traceable, ethical, and safe turns curiosity into loyalty. And loyalty, in a niche market, is everything.

Promoting Sustainability and Ethical Practices

There's also the bigger picture: the environment. Camel milk, by nature, is already more sustainable than cow's milk. Camels need less water. They graze on tougher land. They're gentler on fragile ecosystems. But until now, there's been no easy way to track or prove that impact. Blockchain changes that. It can log water use, transport emissions, energy inputs. It gives producers a chance to document and share their sustainability efforts in a way buyers and regulators can verify.

More than that, blockchain opens the door for incentives. Want to encourage herders to adopt solar-powered cooling or better grazing practices? Build it into the system. Offer bonuses. Record the change. Over time, the tech doesn't just track sustainability—it promotes it. It becomes a driver for better practices across the chain.

And while we're talking about fairness, labor deserves attention too. In parts of the world where camel milk is produced, labor conditions can be murky. Workers often have no contracts, no protections. Blockchain makes it possible to log labor practices, verify wage payments, and ensure that ethical standards are being met. It doesn't solve exploitation on its own—but it makes it a lot harder to hide.

What we're looking at, really, is a shift in how the dairy industry—even its smallest, most niche segments—can function. Blockchain doesn't magically fix broken supply chains. But it shines a light into the places that have gone unseen for too long. It holds people accountable. And it lets others, often for the first time, step into the spotlight with their own stories and value.

For camel milk, this shift is especially important. The product is rich, rare, and rooted in tradition. But it can't grow globally without systems that support fairness, trust, and transparency. Blockchain, if implemented with care, offers a way to build those systems from the ground up.

And in doing so, it doesn't just help camel herders or new markets. It sets a precedent. A way forward. A model for how technology can serve people—not the other way around.

CHALLENGES TO BLOCKCHAIN ADOPTION

While blockchain offers a promising path forward for the dairy industry—especially in small and emerging markets like camel milk—the road to implementation isn't exactly smooth. The technology is powerful, yes, but it doesn't exist in a vacuum. It has to work within real-world conditions: limited resources, fragile infrastructure, fragmented supply chains, and, just as importantly, human hesitation.

Technological Barriers

Adopting blockchain sounds great on paper. In practice, though, it runs into some hard realities. The first is technological capacity. In many places where camel milk is produced, for instance, there's very little digital infrastructure to build on. Internet access is patchy or unreliable. Electricity isn't always stable. And the people doing the production—the herders and small-scale farmers—often haven't used anything more advanced than a feature phone. Expecting them to suddenly engage with sensor networks, digital wallets, or cloud dashboards is unrealistic without serious support.

That's not to say it can't be done. But it takes time, training, and a level of trust-building that goes far beyond rolling out a few devices and explaining how to scan a QR code. Producers need to understand why blockchain matters, how it protects them, and what benefits they can actually expect to see—not just now, but down the line.

Even when the willingness is there, the systems often aren't. A lot of the dairy sector—especially in traditional markets—still runs on manual logs, verbal agreements, and legacy platforms. Trying to layer blockchain on top of that without disrupting the flow is complicated. Getting old systems to "talk" to new ones isn't a quick task, and technical compatibility can be a real sticking point.

Cost and Infrastructure Issues

And then there's the cost. Blockchain systems aren't free. Setting them up means investing in hardware, software, and skilled people to run and maintain everything. Sensors need to be installed and maintained. Platforms need to be updated. Smart contracts don't write themselves. For a large multinational dairy operation, that might be feasible. But for a camel herder trying to keep their animals hydrated in a desert? It's a tough ask.

Even cooperatives or community networks often don't have the capital to take this on alone. Financing models—grants, subsidies, development partnerships—have to be part of the conversation. Otherwise, the very people blockchain is supposed to help will be left behind.

Privacy and Regulatory Concerns

Another issue that's easy to overlook is privacy. Blockchain is praised for its transparency, but in many cases, people don't actually want all their information out in the open. A small producer may not feel comfortable sharing the exact coordinates of their farm, or the full details of their pricing and yields. Larger players might worry about competitors seeing more than they should. And frankly, in some cases, the lack of transparency has been part of how certain intermediaries stay profitable. Exposing inefficiencies or questionable practices can create resistance, not enthusiasm.

So, privacy matters. And so does control. That's why permissioned blockchain systems—where access is limited and data sharing is customizable—are often more appropriate than fully public models. People need to know they're not handing over their data to a black hole or risking exposure.

Regulation is another gray area. In some countries, blockchain is encouraged. In others, it's barely acknowledged in law, or even viewed with suspicion. Add in the complexities of food safety standards, trade certifications, and international compliance rules, and you've got a long list of variables to manage. For something like camel milk, which may cross multiple borders on its way to a consumer, the regulatory load can be intense—and sometimes contradictory.

And it's not just about laws. It's about clarity. Many governments still don't have a clear framework for how blockchain should be used in agriculture or food systems. That uncertainty makes it harder for investors to commit and for small organizations to plan.

On top of all that, there's a more subtle challenge: trust in the technology itself. Not everyone is convinced. In fact, for some stakeholders—particularly those whose businesses rely on opacity or control over information—blockchain can look like a threat. Why support a system that could replace you, or expose inefficiencies you've quietly accepted as normal? That kind of resistance doesn't always show up in meetings, but it can slow down adoption all the same.

And of course, blockchain is only as good as the data it holds. If bad data goes in, bad outcomes come out. If someone lies about milk quality and the system has no checks to verify it, that information becomes permanent. That's why technical solutions need to be backed by strong governance—who enters the data, who checks it, and how errors are resolved.

So, where does that leave us?

The truth is, blockchain isn't a magic fix. It's a tool—a very powerful one—but it has to be handled thoughtfully. Its success depends not just on code or sensors or dashboards, but on people. On local context. On trust, training, funding, and support.

It requires partnerships between governments, technologists, producers, buyers, and consumers—because no one group can carry the whole thing on their own.

If those partnerships come together, though—if the rollout is slow, smart, and sensitive to real-world dynamics—then the obstacles aren't insurmountable. They're just part of the path. And overcoming them isn't just about tech adoption. It's about shaping a food system that's more fair, more resilient, and more honest.

For the camel milk industry especially, where potential and difficulty often go hand in hand, blockchain could help flip the equation—if we're willing to do the work to make it real.

FUTURE PROSPECTS

Looking ahead, blockchain in the dairy world—especially in the case of camel milk—isn't just a technical evolution. It's part of a much broader shift toward re-thinking how food systems can function: more fair, more open, and better equipped to deal with the complexity of today's markets. We're still early in that journey, but it's clear the next wave will be shaped by two things—innovation in the tools themselves, and the ability to bring those tools into the hands of people who need them most.

Innovations to Enhance Blockchain Integration

One of the most exciting things happening now is the way blockchain is starting to talk to other technologies. On its own, blockchain creates trust and traceability. But paired with tools like IoT and AI? That's where things start to scale. Sensors on a milk truck can feed real-time data into a blockchain ledger, and AI can analyze it instantly—predicting spoilage risks, optimizing delivery routes, even helping herders know the best time to milk based on past yield patterns and environmental conditions.

These systems are starting to get smarter, too. The early blockchain platforms were slow, energy-hungry, and hard to scale—especially for sectors like dairy where data flows constantly. But that's changing. New models like Layer 2 chains and green protocols are making blockchain faster and cleaner to run. That matters—not just to keep up with supply chain volume, but to stay aligned with climate goals. No one wants to improve sustainability on one end of the chain only to burn energy on the tech side.

Accessibility is also improving. A few years ago, blockchain tools felt like they were built only for engineers. Today, we're seeing mobile interfaces designed for farmers, voice-enabled dashboards for field workers, and simple apps that let herders check payments, verify buyers, or upload delivery data—all from a basic smartphone.

That kind of interface shift is crucial. If camel herders in remote areas are going to use blockchain, the tools need to speak their language—literally and metaphorically.

Expansion to Emerging Markets

The real power, though, might lie in where blockchain goes next. And that's into places where traditional systems never really worked. In emerging markets, where infrastructure is weak and records are scattered—or nonexistent—blockchain offers a chance to start fresh. You don't need massive server rooms or paper trails. Just a small device, a bit of connectivity, and a platform that's been tailored for the context.

Camel milk producers across the Middle East, North Africa, and parts of Asia could leapfrog the old supply chain model altogether. They don't need to build what industrial dairy producers spent decades creating. With the right system, they can log a batch, verify its quality, trace its journey, and get paid fairly—with the whole transaction recorded and locked in real time. It's not about catching up. It's about doing it differently—and in many ways, better.

This also opens doors to global markets that were previously out of reach. One of the reasons camel milk hasn't gone mainstream is that buyers can't always be sure what they're getting. There are questions about hygiene, authenticity, safety. But blockchain answers those questions, and it does so with data, not marketing. A buyer in Germany or Singapore doesn't have to guess—they can scan a code and see the product's full chain of custody. That's the kind of transparency that moves a product from curiosity to credibility.

Role of Government and Industry Partnerships

And then there's financial inclusion. Right now, many small-scale producers don't have access to loans, insurance, or even bank accounts. But blockchain doesn't just move milk. It moves money—securely, transparently, and in ways that build trust. A camel herder who delivers consistently can build a digital track record. That record becomes a form of credit, a gateway to financing, and a way to grow beyond subsistence.

To make this future real, though, governments and industries need to work together. It's not just about funding. It's about policy, training, and coordination. Regulators have a role to play—setting guidelines that protect data without blocking innovation, and defining standards that help systems connect. Incentives matter too. Grants, tax breaks, and startup support can lower the barrier for producers who want to adopt blockchain but simply can't afford the first step.

Public-private partnerships are key here. A national agricultural ministry working with a blockchain firm and a camel milk export board could design and test a program tailored for local realities—something lightweight, scalable, and designed to show early wins. Once the benefits are visible, the case for scaling becomes much easier to make.

And none of it works without capacity building. People on the ground need to know how to use the tools—and why they matter. That means more than a one-time workshop. It means ongoing digital literacy support, mobile training teams, peer networks, and technical helpdesks that don't vanish after the pilot ends. The technology is only as useful as the people who understand it.

Standards are another piece of the puzzle. Right now, every supply chain experimenting with blockchain seems to be building their own protocols. That's fine at first—but long term, if these systems can't talk to each other, the benefits get stuck in silos. Governments and industry bodies should push for open standards, shared infrastructure, and data formats that allow milk—from cows, goats, camels, or otherwise—to be traced in the same language.

What's ahead isn't just a more efficient supply chain. It's a smarter one. A more transparent one. And, if we get this right, a fairer one. Blockchain won't solve every problem in dairy—but it could solve the ones that have been ignored the longest. It won't replace trust. It'll make trust visible.

For camel milk, especially, this is a rare opportunity. A moment to step onto the world stage not just with a great product, but with a system that proves its value, honors its producers, and meets the standards of a new generation of consumers. With the right tools and the right partnerships, the future of camel milk doesn't just look viable—it looks transformative.

CONCLUSION

In the end, this isn't just about blockchain, or tech, or even milk. It's about people. About building systems that actually reflect the work and worth of the people in them.

The dairy industry, in all its forms, has always depended on invisible hands. Farmers who wake up early to tend animals. Herders walking long distances with milk cans. Small processors doing their best with limited resources. These people keep the whole thing going—but they're rarely seen, and even more rarely heard.

Blockchain can't fix every problem in the dairy world. It won't make a camel give more milk, or replace broken roads with cold chains. But what it can do—what it already does in places where it's being tried—is make the invisible visible. It creates a shared truth. A record that can't be quietly rewritten. It's a quiet but powerful shift.

In the wider dairy sector, this means less waste, better quality, and clearer logistics. It means being able to say, with confidence, where something came from and how it was handled. And that matters—especially when food crosses borders and people want more than just empty labels.

But for camel milk, the stakes are different. This isn't just a supply chain waiting to be digitized. It's a scattered, underdeveloped, often misunderstood sector that's full of potential—and full of barriers too. Low volume. Harsh climates. Limited infrastructure. And yet, despite all that, demand is growing. Slowly, but steadily.

If there's ever been a product that needs a new kind of system to succeed, it's camel milk. And blockchain—if it's adapted properly—might be exactly that system.

Not because it's high-tech. But because it can be honest. Because it can track things that matter: who produced the milk, how it was stored, whether the people involved were treated fairly, whether the land was used wisely. Because it can carry those stories, and make them part of the product's value.

That's not about data. It's about dignity.

And yes, adoption will be slow. Training takes time. Connectivity is uneven. Regulations are catching up. But the direction is good. And the early signs? Promising.

What's ahead isn't just better supply chains. It's a better understanding of who's in the chain, and how they're connected. It's a world where trust isn't just a feeling—it's something you can prove.

So maybe the biggest thing blockchain brings to dairy isn't technology. It's a chance to rebuild trust—between producers and buyers, between farmers and regulators, between traditions and the future.

And that might be just what camel milk needs to grow, and to be seen for what it really is: not just a niche product, but part of a much bigger, more meaningful story.

Chapter 8
The Digitization of the Moroccan Bovine Dairy Sector:
Current Situation, Challenges, and Prospects

Iliass El Ghazrani
https://orcid.org/0009-0003-2532-7747
Institute Agronomic and Veterinary Hassan II, Morocco

Noureddine Belkadi
International Academy of Agricultural Technology and Innovation, Casablanca, Morocco

Abdelillah Araba
Institute Agronomic and Veterinary Hassan II, Morocco

Loubna El Mansouri
Institute Agronomic and Veterinary Hassan II, Morocco

Yassine Mouniane
https://orcid.org/0000-0002-6037-5301
Faculty of Sciences, Ibn Tofail University, Kenitra, Morocco

ABSTRACT

The Moroccan bovine dairy industry is modernizing through digital technologies to enhance management and productivity. This study evaluates digitalization's current state, obstacles, and development opportunities, including the reliability of a 3D camera for automatic body condition scoring of dairy cows. Analysis of

DOI: 10.4018/979-8-3693-9949-1.ch008

farms (15), technology suppliers (4), milk collection centers (9), and processing plants (4) reveals diverse adoption levels: all farms use surveillance cameras, 8 use automatic fans and misters, 5 have automatic weighing systems, 7 monitor mastitis with conductivity meters, and 4 use pedometers. Software like ISALAIT and Dairy Comp 30 support herd management. Digital technology suppliers offer innovative solutions and technical support, while milk collection centers and processing plants focus on traceability and automation. Financial and technical support and ongoing training are essential to accelerate digital transformation in the sector.

INTRODUCTION

The dairy industry in Morocco is also considered of strategic importance to the agricultural sector and economic identity of Morocco, encompassing nearly 96% of national needs for milk and milk derived products. While there are more than 260,000 farms in the dairy sector, it is a significant source of income for a large number of rural families and contributes to maintaining stability in agriculture areas (Maroc Lait, 2022). However, like many agri-food sectors globally, it is facing more challenges than ever before. The increasing competitiveness of markets, product traceability and quality demands necessitate extensive modernization of the practices of the sector, as digitalization is being recognized as a necessary response to these challenges of modernity (Kozina & Semkiv, 2020).

Besides executing tasks with automation, digitalization gives more precise and optimized management of the dairy production process. Digitalization is the link of various digital technologies, including smart sensors, herd management software, the Internet of Things (IoT) and, in some cases, blockchain for full traceability of products (Bahn et al., 2021). These advancements provide effective management of the farm that will allow productivity gains and thereby ensure the constant quality of dairy products (Hostiou et al., 2014). Digitalization can optimize tracking of animal health and monitoring of production processes and in doing so, it can not only reduce costs of operation but increasingly satisfy the sustainability and transparency needs of consumers and international markets (Shingh et al., 2020). The use of such technological advances in the dairy industry has started, but remains inconsistent throughout Morocco. While certain large farms and dairy plants can implement sophisticated data management and quality assurance systems, most small farms and milk collection centers remain far from accessing such tools, due to economic considerations and training (Bahn et al., 2021). These differences demonstrate the serious challenges that still remain to modernize the entire dairy value chain.

Precision breeding is a critical feature of this modernization. Precision breeding addresses the use of new technologies to collect and harness data related to animal health, reproduction, and productivity in real time (Neethirajan & Kemp, 2021). Precision breeding, also relies on the installation of, devices like sensors, novel automated systems, and management software that allows for farmers to more rapidly and intelligently respond (Halachmi et al., 2019). Studies around the world now demonstrate that the ability to use precision breeding technology and monitoring can lead to substantial increases in production and income through reduction of loss and productive resource use. Bewley (2010) showed that precision breeding improved herd management through improved disease detection while allowing for refinement of feeding practices. Product traceability is another key component of digitalization (Xu et al., 2020). Traceability, which is increasingly being ensured by technologies such as block chain technologies, let users not only trace the food supply chain starting from raw milk through to the consumer's point of sale (Tan & Ngan, 2020). This level of transparency is increasingly becoming a crucial part of expectations for consumers for food quality and safety.

Not withstanding the advancements this field is seeing, the digitalization of the industry is not only about technology, it is also about the current support for different actors via training efforts, together the support policy that enables access to digital tools (Misiuk & Zakhodym, 2023). Technology suppliers have a significant role in providing solutions that are adapted to the needs of the farm, as well as providing continuous training opportunities to use that technology optimally.

This study focuses on the state of digitalization in the Moroccan dairy industry, while analyzing the barriers and opportunities for greater and more harmonized digital uptake. The article will follow tangentially the methodology used, the outcomes of the field survey, and advances in recommendations for accelerating the digital transition of this prominent sector.

MATERIALS AND METHODS

Study Context

The research into the digitization of the Moroccan cattle dairy industry was done in six key regions: Fes-Meknes, Rabat-Sale-Kenitra, Tanger-etouan-Al Hoceïma, Souss-Massa, Casablanca-Settat, and Beni Mellal-Khenifra. These regions were selected because they are major dairy producing areas, and there specialist and diversities of farm and herd managements for us to assess the instability in the acceptance of digital technologies due to local infrastructures. The study focused on four main categories of dairy industry players: dairy farms (large and medium-sized), digital technology providers, milk collection centers and processing plants. The choice of these four types of player gives us a complete overview of the value chain, and enables us to analyze the impact of digitalization at all levels of the industry.

Participants and Sampling

Dairy Farms

This research project involved a sample of 15 farms, with 6 being large farms and the remaining 9 being medium-sized farms, which were targeted to identify farms that were likely to adopt digital technologies in their herd management. We wanted to evaluate farms that had at least some established technologies versus farms that did not. Participating farms used did use: video surveillance cameras, animal identification systems using radio frequency, systems that had sensors for reproductive, and animal health management, systems for automated weighing or feeding.

Digital Technology Suppliers

The objective of sampling suppliers of technical tools was to build a representative sample of companies offering state-of-the-art solutions for livestock management, which included reproduction, animal health, herd management, milking parlors and fitting equipment. We were interested in suppliers who were already entrenched on the farm level, as they would be best aware of the specific dealings of the Moroccan dairy sector. In the end, we were able to survey 4 companies and collect relevant information on available technologies, regarding adoption of their technology by farmers, as well as development and collaborative opportunities for the digitalization of the dairy sector.

Milk Collection Centers

Nine milk collection centers were included in the study, located in Morocco's main dairy zones. These centers play a crucial role in milk quality management, collecting, storing and transporting raw milk to processing plants. Technologies evaluated at the collection centers include: Eko Milk and Lactoscan systems for milk quality analysis (fat, density, temperature), digital collection data management systems and partial traceability software. The centers were selected according to their size and their use of digital technologies for milk management. The largest centers were prioritized for the study because of their central role in the dairy supply chain.

Processing Plants

Finally, four milk processing plants were selected to assess the integration of digital technologies in processing and distribution. These plants use tools such as Milkoscans to analyze milk composition, ERP (Enterprise Resource Planning) systems to manage production flows and centralize data and route management tools such as MuuMap and Ortec to optimize logistics. Selected plants are those that have already begun to adopt advanced traceability systems and are in the process of developing digital platforms to improve supply chain management.

Tools and Technologies Used

Data Collection Tools

Study data were collected via structured questionnaires and semi-structured interviews with farm managers, technology suppliers, collection center operators and processing plant managers. The questionnaires were designed to assess: The level of adoption of digital technologies, Users' perceptions of the advantages and disadvantages of these technologies, Obstacles encountered in integrating digital tools. Semi-structured interviews were used to gather more in-depth qualitative information on the technical difficulties encountered by operators, as well as on the solutions envisaged to improve technology adoption.

Questionnaire Headings and Themes

For farms, sections included general information (location, farm structure, production level), RFID identification technology and its use (automated weighing, animal sorting), intelligent equipment for livestock buildings (surveillance cameras, automatic climate control systems, self-triggering fans and misters), as well as devices for monitoring reproduction, pathologies and body condition score using 3D imaging. Other sections explored management software, technology adoption challenges and development prospects.

Milk collection centers and processing plants were also surveyed on similar themes, with a particular focus on traceability technologies, process automation and milk quality management. Questions addressed processing capacity, digital tools used for milk collection and analysis, and development prospects in these sectors.

Questionnaires for suppliers covered topics such as proposed technological solutions, technical support offered to farms, challenges encountered in the process of technology adoption by users, as well as development prospects for future innovations in the sector. In addition, particular emphasis was placed on the top 3 equipment products marketed by each supplier, enabling us to understand which solutions are preferred on the market and their potential for adoption by farms.

Complementing these surveys, a specific study was carried out to assess the reliability of a 3D camera used in automated body condition scoring of dairy cows. This technology was tested on a farm equipped with it, with a sample of 63 cows spread over three stages of lactation: early, middle and late. The results obtained by the 3D camera were compared with manual evaluations carried out in collaboration with the farm's zootechnician, enabling the effectiveness of this automated method to be validated.

Data Analysis Approach

Quantitative Analysis

The results of the questionnaires were compiled and analyzed using descriptive statistics. Frequency tables and comparative graphs were used to visualize differences in technology adoption between large and medium-sized farms, technology suppliers, as well as milk collection centers and processing plants. The main variables studied include: The rate of adoption of different technologies (RFID, sensors, smart cameras, ERP, etc.), Perceived benefits in terms of productivity, milk quality, and resource management, as well as financial and technical barriers encountered.

To assess the concordance between body condition measurements obtained by visual assessment in the field and those provided by the 3D camera, SPSS software was used to perform statistical analyses.

Correlation. Pearson's correlation was calculated between field scoring data and 3D camera data for each lactation phase. This analysis determined whether the two types of measurement followed a similar direction.

Mean Comparison Test. To go beyond simple correlation, a paired-samples Student's t-test was performed. This test was used to assess whether there were any significant differences between the measurements obtained by the 3D camera and those taken in the field. A p-value of less than 0.05 was considered indicative of significant differences, confirming the importance of accurate body condition measurements.

Qualitative Analysis

The data from the semi-structured interviews were subjected to a content analysis to identify the main obstacles and opportunities linked to the adoption of digital technologies. This analysis identified recurring trends in farmers' perceptions of the usefulness of technologies and the difficulties associated with their implementation.

RESULTS AND ANALYSIS

Technology Adoption on Dairy Farms

Using RFID Systems to Identify Cows

The results show that all the farms surveyed have adopted the radio frequency identification (RFID) tags recommended by the ONSSA, particularly following the launch of the National Identification and Traceability System (SNIT) in 2015. These tags are used for a variety of purposes, ranging from parlour monitoring to automated weighing and animal sorting. The widespread adoption of RFID systems is evidence of a successful first phase of digitisation on dairy farms in Morocco. However, some farms have reported maintenance problems and frequent breakdowns of RFID readers, which limits the effectiveness of these systems over time. This technical problem is also exacerbated by software incompatibilities between the equipment and the data management systems, making it difficult to integrate the data collected into herd management software.

Farms therefore need to invest in regular maintenance and improvement of digital infrastructures to ensure continuous and reliable traceability. In addition, the integration of RFID readers into more advanced herd management systems would be beneficial to prevent the loss of critical data and optimise the use of these devices in the long term.

Adoption of Intelligent Building Management Equipment

Intelligent technologies such as surveillance cameras, fans and automatic foggers are increasingly present on the farms studied, particularly on large farms. These systems aim to optimise animal welfare conditions and improve the day-to-day management of livestock farms. For example, automatic foggers and fans reduce heat stress for animals, particularly in hot regions such as Souss-Massa and Béni Mellal-Khénifra. The results reveal a significant disparity between large and medium-sized farms. Some medium-sized farms have actually adopted these technologies, while others are less inclined to do so because of the high installation and maintenance costs (Figure 1 & 2).

Larger farms generally have the resources to invest in automatic weighing systems and environmental control devices (temperature and humidity), which enable optimised management of animal health and feeding. On the other hand, most medium-sized farms feel that this equipment is better suited to large herds and does not justify the investment for their size. This illustrates a dynamic where large farms have an advantage in terms of modernisation and adoption of advanced technologies, while the majority of medium-sized farms still face significant financial barriers.

Figure 1. Installation of intelligent equipment for livestock buildings on the large farms surveyed

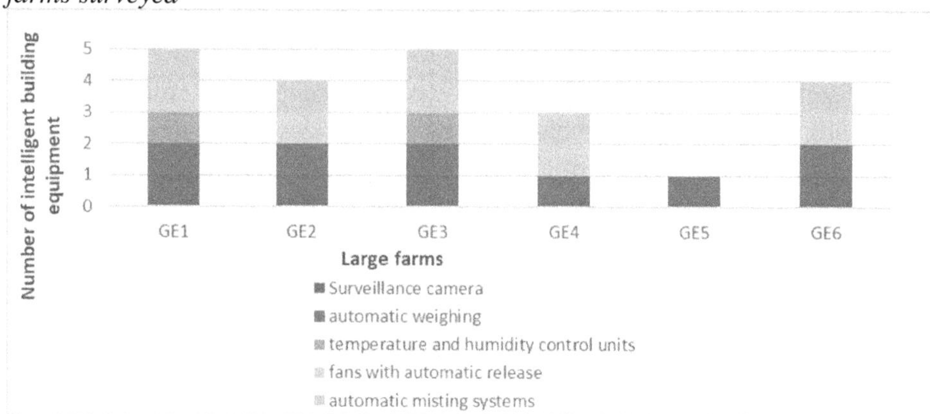

Figure 2. Installation of intelligent barn equipment on medium-sized farms surveyed

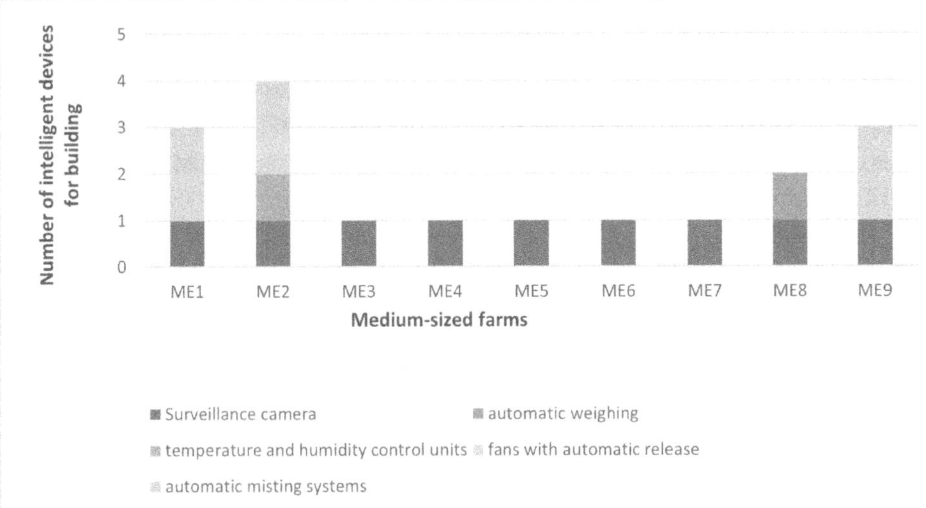

Monitoring Reproduction and Behaviour

Monitoring reproduction and feeding behaviour is a fundamental aspect of precision breeding. Of the six technologies examined in the survey, pedometers and accelerometers stand out as the tools most frequently used for heat detection. In fact, several farms use the pedometer, while only one uses the accelerometer to manage reproduction. However, some farms decided to stop using accelerometers, citing their perceived inefficiency and the frequency of false alarms, which ultimately increased rather than decreased the workload for farmers. This highlights the need to improve the reliability of these technologies and ensure adequate training for farmers.

As far as calving monitoring is concerned, the majority of farms continue to rely on manual methods or conventional video surveillance systems to detect signs of calving. The general absence of automated calving detection systems, even on large farms, indicates that these technologies are under-utilised, despite their potential benefits in reducing perinatal mortality and improving animal welfare. However, one farm has adopted an accelerometer for calving monitoring, reporting a significant improvement in the early detection of calving complications.

In terms of feeding behaviour, two medium-sized farms use accelerometers to monitor rumination and chewing activity. One farm applies them during the transition phase, focusing on monitoring feeding behaviour to prevent metabolic disease, while another uses them continuously to monitor all cows. This illustrates the commitment to taking advantage of available technologies to improve animal management.

Animal Health Monitoring: Detection of Lameness and Mastitis

The results show that the pedometer is one of the most popular devices for detecting lameness, making it possible to detect abnormalities in the gait of cows and intervene quickly in the event of a problem. However, other devices such as pressure plates or cameras for image analysis are still little known or used on Moroccan farms. Mastitis detection technologies, in particular conductivity meters that measure the electrical conductivity of milk, have been widely adopted on farms, with positive results in terms of saving time and reducing economic losses due to subclinical mastitis. This shows that farmers are aware of the importance of these technologies for the management of common diseases, although their adoption remains limited by high initial costs and maintenance problems

Body Condition Assessment

The results of this study show that the DeLaval 3D camera is globally reliable for body condition assessment of dairy cows, with strong correlations observed early (0.71), mid (0.86) and late lactation (0.78). These correlations indicate an overall consistency with the manual estimates. However, at the beginning of lactation, where the BCS is often between 2 and 3, significant differences were noted ($p = 0.04$) (Table 1), suggesting that the accuracy of the device is less at this stage. This limitation could be due to a number of causes, including insufficient lighting and inappropriate calibration of the DeLaval 3D camera for this stage, which make it difficult to accurately capture measurements assessing cow body condition.

Table 1. Results of 3D camera evaluation for body condition scoring of dairy cows

Lactation phase	**Correlation Analysis**	**Student's t-test**	
		Average difference(d)	**p-value**
Start of lactation	0,71	0,19	0,04
Mid-lactation	0,86	0,04	0,36
End of lactation	0,78	0,09	0,07

Technologies Marketed by Digital Solution Providers

The survey of digital technology suppliers to the beef dairy industry has identified a number of PLCs and sensors that are essential for farms. Among the equipment most commonly sold are automatic fans and misters, environmental control units, automatic scrapers, robotic milking systems, automatic milk dispensers (AMDs) and automatic concentrate dispensers (ACDs) (Figure 3). This equipment is increasingly being incorporated into large Moroccan farms to improve animal welfare and productivity in the face of climatic constraints.

Sensors are also crucial for proactive herd management. The survey revealed the marketing of pedometers, accelerometers, thermal cameras, as well as force sensors and pressure plates for lameness detection. The pedometer is the most commonly used, followed by the accelerometer (Figure 4). These sensors make it possible to monitor the physical activity and health of cows in real time, facilitating early detection of problems such as heat or locomotion disorders.

Figure 3. PLCs marketed by the digital technology suppliers surveyed

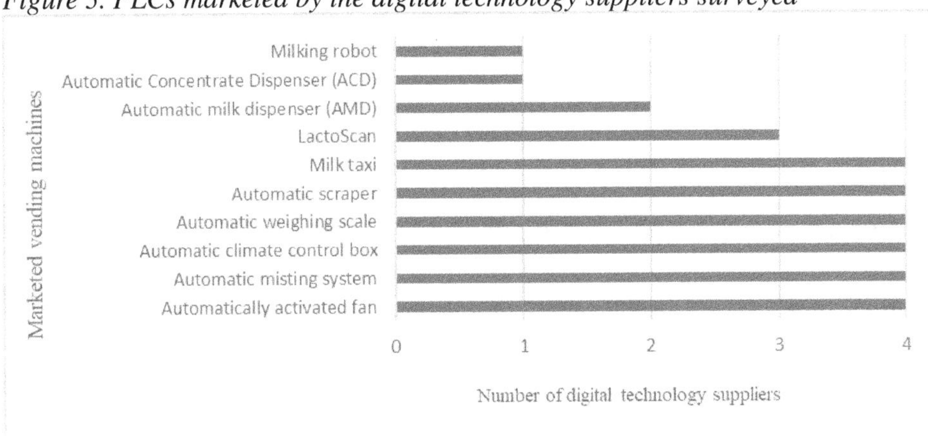

Figure 4. Sensors marketed by the digital technology suppliers surveyed

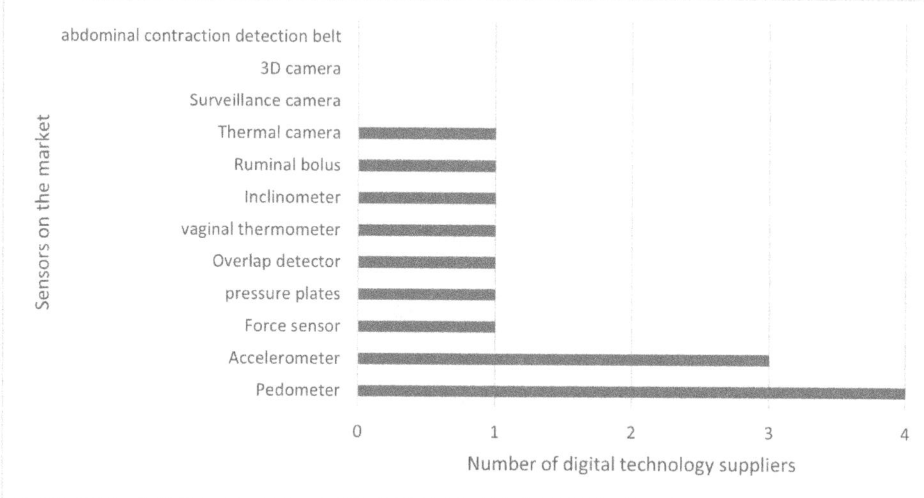

Adoption of Technologies by Milk Collection Centers

Milk collection centres play a crucial role in the quality management and traceability of dairy products. The results show that most collection centres use milk quality analysis technologies such as the Lactoscan and Eko Milk systems, which mainly measure the fat content, density and temperature of milk. However, a number of challenges remain in terms of equipment uniformity, making it difficult to integrate data between different centres and dairy plants. In addition, some centres report difficulties in integrating digital systems due to a lack of adequate infrastructure, such as stable internet connections in rural areas.

Digitised collection centres are reporting significant improvements in milk quality management and product traceability, but these technologies have not yet been rolled out to all cooperatives. Investment in infrastructure (electricity, internet) and in operator training are essential if these technologies are to be adopted more widely throughout the country. In addition, subsidy programmes or low-interest loans could encourage collection centres to adopt these technologies and thus improve the overall efficiency of the sector.

Digitization Technologies in Processing Plants

Milk processing plants have integrated digital technologies to improve both milk quality and product traceability. Among these technologies, the use of tools such as Milkoscans makes it possible to obtain precise data on the composition of milk, making it easier to manage the processing and distribution processes. However, integration challenges remain, particularly with regard to the interconnection of systems between the various players in the sector, including collection centres, factories and distributors. Although ERP systems are used to centralise production data, the lack of a centralised national platform limits the fluidity of traceability throughout the value chain. What's more, milk transport is a crucial stage in the dairy supply chain. At this stage, real-time tracking and route optimisation are essential to ensure the quality of the milk until it reaches the factory. GPS systems are used to track collection trucks, enabling routes to be monitored, delays to be prevented and vehicle management to be optimised. The integration of route optimisation software, which exploits massive data and artificial intelligence techniques, also helps to improve logistics efficiency.

To guarantee product safety and traceability, sensors are deployed on the trucks to detect any unauthorised opening during transport, strengthening the fight against potential adulteration. These technological advances are designed not only to improve collection efficiency, but also to ensure strict traceability throughout the dairy value chain. Using blockchain for full and transparent traceability is a vision for the future with added consumer confidence. By centralizing data on a tamper-proof blockchain, blockchain could guarantee the quality and origin of dairy products throughout the supply chain, from the farm to the consumer. Nonetheless, this technology is still seen in a new era in Morocco and will require even more investments and collaboration across sectors to arrive at full implementation.

Economic and Productive Impact of Digitalization

The uptake of digital technologies has a direct effect on dairy farm productivity and the economic profitability of the sector. Digital technologies such as sensors, herd management systems and automated monitoring tools contribute to better animal health management, lower operational costs, and increased productivity. On farms with automated monitoring systems and mastitis detection systems, farmers have reported a considerable reduction in losses associated with diseases, like mastitis, which is one of the main drivers of economic losses in the dairy sector. Herd management systems such as ISALAIT, Dairy Comp 30 and Bovisync allow farmers to study and evaluate animal health and their productive abilities in great detail, while also optimising feeding, reproductive performance and milk production. These technologies also allow predictive analyses that help farmers anticipate problems and then make decisions from reliable data. For example, from using the automatic body condition prediction (NEC) feed tools with 3D cameras you can improve the nutrition monitoring of cows through diet adjustment supporting milk productivity according to the needs of the animals. The economic advantages of digitisation are demonstrated through a notable increase in productivity per cow. Farms which use digitisation fully average around 10,000-11,000 litres of milk per lactation, compared to about 7,500 litres for farms not using digitisation. This difference cannot be attributed to digitisation, but is an important element in managing resources, allowing increased production, and overseeing manual management and human error costs.

Challenges and Limitations to Technology Adoption

Although digitalisation offers unquestionable advantages, a number of barriers hinder it. The major challenges defined in this study are affordability, lack of training, weak rural infrastructure and willingness to change.

Technology Costs and Affordability

The total cost of adopting digital technologies limits the utilization of digital technologies by farms, especially those that are small or medium-sized. For example, installing sensors or 3D cameras in a barn requires a large initial capital outlay which in many cases small farms cannot afford. This investment represents the cost of purchasing the sensors and cameras, the maintenance of the equipment, the software updates, and the productivity lost during installation – significant impediments to wider adoption of green technologies. Medium-sized farms in particular express doubts around the business case for going digital.

Training and Technical Support

Farmer and manager training is the most important part of successfully utilising these technologies. While results indicate that farms are beginning to adopt the technologies, they are struggling to incorporate them into day-to-day management activities, primarily because of insufficient technical skills. Farmer dissatisfaction was also related to suppliers not providing sufficient follow-up and/or technical support once the equipment was installed. To address this, we strongly recommend developing ongoing training programmes which help farmers become proficient with the digital tools available to them.

Digital Infrastructure and Connectivity

Infrastructure challenges are another significant barrier to digitalisation. Many farms, especially those in rural and remote areas, are not yet equipped with adequate internet connection, which hinders their capacity to use connected management systems or access data in real-time. Also, intermittent power supplies in remote and rural regions limits the use of automated technologies, such as smart sensors and remote monitoring. Improvements to rural infrastructure, specifically internet connectivity and access to consistent power, are also necessary so that digital technologies can be broadly adopted throughout the sector.

Resistance to Change and Selective Adoption

Resistance to change is also a major barrier. Many farmers, especially those who run medium-sized farms, simply want to keep doing what they have always done. They believe that adopting the new technologies will create more complex tasks and labor - instead of easier tasks and/or less labor. Often, this resistance is connected to the overall lack of information on the economic and productive contributions of digital technologies. More information and practical examples of precision breeding technologies can help to overcome this resistance.

Prospects for Improvement and Development Opportunities

Despite the challenges, the digitalisation of the dairy sector in Morocco presents considerable opportunities for modernising production and improving the sector's competitiveness. This study identifies several areas for development.

National Data Centralisation Projects

Simalait will start in January 2024 with the purpose of nationalising data on animal identification, insemination and milk recording. This project will enhance coordination of industry players and allows for digital technology integration. It will also provide centralized data which can enhance herd management; with enhanced analysis tools for more precise output optimisation and assurance of traceability of milk in the supply chain.

Financial Incentives and Subsidies

To promote the adoption of digital technologies, some financial support by public subsidies and low interest loans could be made available to medium-sized farms. This form of financial support would make acquisition of technology a more democratic feat, especially for small farms representing a majority of the total dairy production in Morocco. Public-private partnerships could also support the up-take of technological innovations, which would be supported on the farm and at the milk collection centre.

Training and Skills Development

Another key lever to accelerate the digitisation of the sector is the development of training courses specifically for farmers, managers and technicians. These courses would be offered in relation to herd management systems, smart sensors and automated monitoring systems. This will enable farmers to better appreciate the advantages of digital technologies and to solve the technical problems that farmers face on adoption.

Integration of Emerging Technologies: Blockchain and Artificial Intelligence

The emergence of new technologies and innovations, including blockchain and artificial intelligence (AI), certainly provide an exciting opportunity for the sector's future. Blockchain has the potential to ensure that products are equally transparent and traceable from producer to consumer, while also minimizing the impact of fraud as well as enhancing consumer trust in their local products. AI could be used to sort and analyze large datasets, provide predictive analyses on their animals' health, feed optimization, and dairy performance management.

DISCUSSION

The results of this study highlight several key points concerning the digitalisation of the Moroccan dairy sector. Although digitalisation is seen as an essential lever for improving the productivity and management of dairy farms and other players in the sector, its adoption remains uneven, reflecting disparities in terms of economic resources and infrastructure.

Inequalities in Technological Adoption: A Risk of Widening Disparities

Digitalisation appears to have initially benefitted larger farms that are implementing technologies such as automated monitoring systems and integrated management software. A few medium-sized farms are adopting technologies such as pedometers, accelerometers, and movement sensors; however the majority of medium-sized farms do not have access to these technologies. This highlighted a dichotomy where larger farms are benefiting from these innovations, while some medium-sized farms are slowly evolving, but not as quickly (Hostiou et al., 2014). It's also noteworthy that more than some of the larger farms have not completely adopted these technologies. This illustrates that uptake of digital technologies happens on the basis of the sociocultural infrastructure of a farm, not just the size of the farm, and whether there is the willingness and resources to adopt such technologies. This exemplifies a pattern seen in other agricultural sectors where larger businesses gain from the innovations immanent to smaller farms (Hostiou et al., 2017). This situation raises doubts about the equity of agricultural modernisation in Morocco. If smallholders and medium farms cannot use these tools, the gap could widen in terms of technology, meaning the smallest structures will become less competitive. That is why, in order to democratise access to digital technologies, targeted intervention will be needed, either through public subsidies or new financing methods (Adenuga et al., 2020).

DeLaval 3D Camera: A Promising Tool That Needs Fine-Tuning

The application of the DeLaval 3D camera for body condition scoring (BCS) of cows yielded quite interesting results particularly for cows evaluated in the mid and late part of lactation. The results support studies of Mullins et al. (2019) which found optimal body scores between 3.00 and 3.75. However, some modifications are needed to improve the level of accuracy of the measurements at the beginning of lactation. This indicates that while technology can be effective, it needs to continue to adapt to meet the specific problems of farmers.

Systems Integration: Collection Centers and Plants

Milk collection centres and processing facilities have accomplished strides in utilising technology for the differentiation of raw milk quality and for improving transport logistics. Technologies like Milkoscan and Lactoscan are being used to determine milk composition and ensure milk quality prior to processing. This helps to identify quality issues promptly, which is important in meeting market compliance. At the same time, GPS and transport management systems now allow collection transport to be optimised, leading to reduced delays and reduced quality devaluation at transport. Clearly, these enhanced logistics processes are necessary to achieve efficiency and maintain product quality throughout the logistics of the raw milk supply chain (Karouani & Elgarej, 2021). Additionally, the use of ERP (Enterprise Resource Planning) systems at those plants, aids in operational management, because they consolidate production, quality and logistics data and therefore allow industry players to make decisions on-the-floor based on real-time data which leads to better resource-management. However, although it is hopeful progress, there is definitely room for improvement to achieve complete integration across the national space.

The Role of Suppliers: An Essential Catalyst

Technology providers are essential parts of the digital transformation of farms, although their contributions are limited by licensing restrictions and domestic bureaucracy relating to technology imports. These restrictions can limit access to the innovation needed to modernise farms (Banhazi et al., 2011). The absence of rigorous technical follow-up after the sale of equipment indicates a structural weakness in the way end users are supported and sometimes unable to realize the benefits of the technologies purchased from lack of ongoing training or technical support. Digitisation should be considered as not only an equipment process, but as a supporting process – and should include not only skills transfer but also continuous updates on systems (Halachmi et al., 2019). This evidence is supported by Bianchi et al (2022), which indicates that new technology use without training results in inefficiencies. This can be especially pertinent in rural areas that may have limited access to specific technical services. As a result, training and technical support should be included as an important aspect of the digitalisation strategy.

Prospects for Modernization and Integration of Emerging Technologies

This study's results also demonstrate the untapped potential of emerging technologies such as artificial intelligence (AI) or blockchain. In Morocco, new technologies are still on the first steps of their evolution; however, they represent interesting potentials to enhance overall traceability and productivity in dairy farming. AI could be employed to analyse sensor data to detect early warnings of potential health issues in animals, while blockchain would guarantee complete transparency in the supply chain from input to sales (Xu et al., 2020).

That being said, in order to fully integrate new technology, it is essential to invest significantly in training and building infrastructure. Like many technology use in originating and supply chains, deploying innovations as such innovation depends on the strong co-operation of both the public sector and private parties in developing an environment that is supportive of innovation, while developing systems to ensure that small and medium sized farms are not locked out of the modernisation of the dairy industry.

Prospects for Modernization and Integration of Emerging Technologies

There are some limits to this study that should be considered when shaping future research. First, while the number of 15 farms and 9 collection centres made for a reasonably representative sample, the addition of more small-scale actors would have improved clarity on the dynamics at work throughout the entire sector. Second, variation in technologies used made it challenging to allow for direct comparisons between farms. In this regard, it would be wise to promote the harmonisation of digital tools and sharing of best practices to facilitate effective adoption across the sector. Their farms also were unable to report back all the desired data related to their digital management processes because the information was too sensitive. In order to improve future research, it would make sense to develop a common framework for data sharing amongst the actors in the sector, but also respecting confidentiality obligations. Lastly, our economic analysis is incomplete, particularly with regard to a detailed analysis of the cost-benefit analysis of the technologies adopted. To develop a fuller economic analysis would help inform cases for investment or justification to adopt innovative tools like the 3D camera.

CONCLUSION

This study highlights the current state of digitisation in the Moroccan dairy sector, revealing inequalities in adoption between large and medium-sized farms. Although some large farms have integrated advanced technologies, many medium-sized and small farms are still lagging behind. Collection centres and processing plants benefit from quality tools such as Milkoscan and Lactoscan, as well as GPS systems and software to optimise transport, but improvements are needed for full integration. The role of suppliers is crucial, but administrative obstacles and a lack of technical support limit their impact. Digitisation must be seen as a process of accompaniment, including training and technical support. Appropriate infrastructure in rural areas is also essential to support this transition. To encourage uniform adoption of digitisation, it is essential to develop appropriate public policies, innovative financing solutions and encourage collaboration between industry players. An integrated approach will strengthen the competitiveness and sustainability of the Moroccan dairy industry.

ACKNOWLEDGMENT

We would like to express our gratitude to the Pôle Digital of the Ministry of Agriculture for their collaboration on this study. Their logistical and administrative support enabled us to make progress in our research into the digitalisation of the Moroccan dairy industry.

We would also like to thank all the people and organisations who contributed to this study with their input, data and time.

REFERENCES

Adenuga, A. H., Jack, C., Olagunju, K. O., & Ashfield, A. (2020). Economic Viability of Adoption of Automated Oestrus Detection Technologies on Dairy Farms : A Review. *Animals (Basel)*, *10*(7), 1241. DOI: 10.3390/ani10071241 PMID: 32708279

Bahn, R. A., Yehya, A. A. K., & Zurayk, R. (2021). Digitalization for Sustainable Agri-Food Systems : Potential, Status, and Risks for the MENA Region. *Sustainability (Basel)*, *13*(6), 3223. DOI: 10.3390/su13063223

Banhazi, T. M., Lewis, B., & Tscharke, M. (2011). The Development and Commercialisation Aspects of a Practical Feed Intake Measurement Instrumentation to be Used in Livestock Buildings. *Australian Journal of Multi-Disciplinary Engineering*, *8*(2), 131–138. DOI: 10.1080/14488388.2011.11464832

Bewley, J. (2010). Precision Dairy Farming : Advanced Analysis Solutions for Future Profitability. Academic Press.

Bianchi, M. C., Bava, L., Sandrucci, A., Tangorra, F. M., Tamburini, A., Gislon, G., & Zucali, M. (2022). Diffusion of precision livestock farming technologies in dairy cattle farms. *Animal*, *16*(11), 100650. DOI: 10.1016/j.animal.2022.100650 PMID: 36220000

Halachmi, I., Guarino, M., Bewley, J., & Pastell, M. (2019). Smart Animal Agriculture : Application of Real-Time Sensors to Improve Animal Well-Being and Production. *Annual Review of Animal Biosciences*, *7*(1), 403–425. DOI: 10.1146/annurev-animal-020518-114851 PMID: 30485756

Hostiou, N., Allain, C., Chauvat, S., Turlot, A., Pineau, C., & Fagon, J. (2014). L'élevage de précision : Quelles conséquences pour le travail des éleveurs? *INRA Productions Animales*, *27*(2), 113–122. DOI: 10.20870/productions-animales.2014.27.2.3059

Hostiou, N., Fagon, J., Chauvat, S., Turlot, A., Kling-Eveillard, F., Boivin, X., & Allain, C. (2017). Impact of precision livestock farming on work and human- animal interactions on dairy farms. A review. *Biotechnologie, Agronomie, Société et Environnement*, 268–275. Advance online publication. DOI: 10.25518/1780-4507.13706

Karouani, Y., & Elgarej, M. (2021). Milk-Run Collection Monitoring System Using the Internet of Things Based on Swarm Intelligence. *International Journal of Information Systems and Supply Chain Management, 15*(3), 1–17. DOI: 10.4018/IJISSCM.290018

Kozina, A. M., & Semkiv, L. P. (2020). Sustainable development of dairy farming through the use of digital technologies. *IOP Conference Series. Earth and Environmental Science, 613*(1), 012061. DOI: 10.1088/1755-1315/613/1/012061

Maroc Lait. (2022). maroclait. https://maroclait.ma/fr/stat.html

Misiuk, M., & Zakhodym, M. (2023). Digitization as a tool for revitalizing the livestock industry. *Ekonomika APK, 30*(4), 10–24. DOI: 10.32317/2221-1055.202304010

Mullins, I. L., Truman, C. M., Campler, M. R., Bewley, J. M., & Costa, J. H. C. (2019). Validation of a Commercial Automated Body Condition Scoring System on a Commercial Dairy Farm. *Animals (Basel), 9*(6), 287. DOI: 10.3390/ani9060287 PMID: 31146374

Neethirajan, S., & Kemp, B. (2021). Digital Livestock Farming. *Sensing and Bio-Sensing Research, 32*, 100408. DOI: 10.1016/j.sbsr.2021.100408

Shingh, S., Kamalvanshi, V., Ghimire, S., & Basyal, S. (2020). Dairy Supply Chain System Based on Blockchain Technology. Asian Journal of Economics, Business and Accounting, 13-19. DOI: 10.9734/ajeba/2020/v14i230189

Tan, A., & Ngan, P. T. (2020). A proposed framework model for dairy supply chain traceability. *Sustainable Futures : An Applied Journal of Technology, Environment and Society, 2*, 100034. DOI: 10.1016/j.sftr.2020.100034

Xu, J., Guo, S., Xie, D., & Yan, Y. (2020). Blockchain : A new safeguard for agri-foods. *Artificial Intelligence in Agriculture, 4*, 153–161. DOI: 10.1016/j.aiia.2020.08.002

KEY TERMS AND DEFINITIONS

3D Camera: Developed for dairy farming, this three-dimensional imaging device can evaluate the body condition of cows automatically and objectively, improving nutritional and health decisions.

Blockchain: A secure, transparent, and immutable distributed ledger technology that stores data and can provide traceability and trust in agri-food supply chains.

Body Condition Scoring (BCS): Assessment of the body condition of dairy animals or livestock animals, normally based on a scale of 1-5, that is used to optimize nutrition, animal reproduction and animal welfare. 3D cameras are now being used to define this assessment.

Digitalization: The process of using digital technologies as part of production, management, and distribution processes to improve efficiency, traceability, and sustainability in agri-food systems including the dairy industry.

ERP (Enterprise Resource Planning): An integrated software that supports the management of all business processes, including production, logistics, human resources, and finance by consolidating all real-time data.

Precision Breeding: A technique of livestock reproduction that requires new advanced technologies, such as sensors and management software, to collect and analyze real-time information about each animal's health, fertility, and performance.

RFID (Radio Frequency Identification): An automatic identification technology that uses radio waves to read and record data on an electronic tag or tags attached to individual animals or groups of animals, or equipment such as milking systems with attached milking robots that helps facilitate herd management and traceability.

Smart Sensors: Devices that can collect, analyze, and transmit data on biological or environmental variables to allow farmers to monitor animal health, behavior, or the environment in which the animals live.

Traceability: The ability to track the histories, uses, or locations of a product through the supply chain, from farm to customer; this can often be accomplished through technology, such as blockchain.

Compilation of References

Abdalla, A. K., Ayyash, M. M., Olaimat, A. N., Osaili, T. M., Al-Nabulsi, A. A., Shah, N. P., & Holley, R. (2021). Exopolysaccharides as antimicrobial agents: Mechanism and spectrum of activity. *Frontiers in Microbiology*, *12*, 664395. DOI: 10.3389/fmicb.2021.664395 PMID: 34093478

Abdel-Basset, M., Hawash, H., Moustafa, N., Razzak, I., & Abd Elfattah, M. (2024). Privacy-preserved learning from non-i.i.d data in fog-assisted IoT: A federated learning approach. *Digital Communications and Networks*, *10*(2), 404–415. DOI: 10.1016/j.dcan.2022.12.013

Abourraja, M. N., Kringos, N., & Meijer, S. (2022). Exploiting simulation model potential in investigating handling capacity of Ro-Ro terminals: The case study of Norvik seaport. *Simulation Modelling Practice and Theory*, *117*, 102513. DOI: 10.1016/j.simpat.2022.102513

Adenuga, A. H., Jack, C., Olagunju, K. O., & Ashfield, A. (2020). Economic Viability of Adoption of Automated Oestrus Detection Technologies on Dairy Farms : A Review. *Animals (Basel)*, *10*(7), 1241. DOI: 10.3390/ani10071241 PMID: 32708279

Afzal, A., Mahmood, M. S., Hussain, I., & Akhtar, M. (2011). Adulteration and microbiological quality of milk (a review). *Pakistan Journal of Nutrition*, *10*(12), 1195–1202. DOI: 10.3923/pjn.2011.1195.1202

Ahmad, T., Aadil, R. M., Ahmed, H., Rahman, U., Soares, B. C. V., Souza, S. L. Q., Pimentel, T. C., Scudino, H., Guimarães, J. T., Esmerino, E. A., Freitas, M. Q., Almada, R. B., Vendramel, S. M. R., Silva, M. C., & Cruz, A. G. (2019). Treatment and utilization of dairy industrial waste: A review. *Trends in Food Science & Technology*, *88*, 361–372. DOI: 10.1016/j.tifs.2019.04.003

Akan, G., & Oner, E. T. (2022). Gel properties of microbial polysaccharides. In *Polysaccharides of Microbial Origin: Biomedical Applications* (pp. 579-598). Springer. DOI: 10.1007/978-3-030-42215-8_34

Akhigbe, B., Munir, K., Akinadé, O., Akanbi, L., & Oyedele, L. (2021). IoT Technologies for Livestock Management: A Review of Present Status, Opportunities, and Future Trends. *Big Data and Cognitive Computing*, *5*(1), 10. DOI: 10.3390/bdcc5010010

Al Sidawi, R., Urushadze, T., & Ploeger, A. (2021). Factors and Components Affecting Dairy Smallholder Farmers and the Local Value Chain— Kvemo Kartli as an Example. In *Sustainability* (Vol. 13, Issue 10). DOI: 10.3390/su13105749

Alipio, M., & Villena, M. L. (2023). Intelligent wearable devices and biosensors for monitoring cattle health conditions: A review and classification. *Smart Health (Amsterdam, Netherlands)*, *27*, 100369. DOI: 10.1016/j.smhl.2022.100369

Allioui, H., & Mourdi, Y. (2023). Exploring the Full Potentials of IoT for Better Financial Growth and Stability: A Comprehensive Survey. In *Sensors* (Vol. 23, Issue 19). DOI: 10.3390/s23198015

Almaiah, M. A., Yelisetti, S., Arya, L., Christopher, N. K. B., Kaliappan, K., Vellaisamy, P., & Alkdour, T. (2023). A novel approach for improving the security of IoT–medical data systems using an enhanced dynamic Bayesian network. *Electronics (Basel)*, *12*(20), 1–15. DOI: 10.3390/electronics12204316

Alonso, R. S., Sittón-Candanedo, I., García, Ó., Prieto, J., & Rodríguez-González, S. (2020). An intelligent Edge-IoT platform for monitoring livestock and crops in a dairy farming scenario. *Ad Hoc Networks*, *98*, 102047. DOI: 10.1016/j.adhoc.2019.102047

Amul Dairy Cooperative. (2022). *IoT implementation report: Efficiency and maintenance*. Internal Publication.

Andrew, M., & Jayaraman, G. (2020). Structural features of microbial exopolysaccharides in relation to their antioxidant activity. *Carbohydrate Research*, *487*, 107881. DOI: 10.1016/j.carres.2019.107881 PMID: 31805426

Anedda, R., Melis, R., & Curti, E. (2021). Quality Control in Fiore Sardo PDO Cheese: Detection of Heat Treatment Application and Production Chain by MRI Relaxometry and Image Analysis. *Dairy*, *2*(2), 270–287. DOI: 10.3390/dairy2020023

Angelin, J., & Kavitha, M. (2020). Exopolysaccharides from probiotic bacteria and their health potential. *International Journal of Biological Macromolecules*, *162*, 853–865. DOI: 10.1016/j.ijbiomac.2020.06.190 PMID: 32585269

Anumudu, C. K., Miri, T., & Onyeaka, H. (2024). Multifunctional Applications of Lactic Acid Bacteria: Enhancing Safety, Quality, and Nutritional Value in Foods and Fermented Beverages. *Foods*, *13*(23), 3714. DOI: 10.3390/foods13233714 PMID: 39682785

Aparnathi, K. D., Shaikh, A. I., & Patel, S. I. (2020). *Qualitative Tests for Detection of Common Adulteration in Milk. Directorate of Research Anand Agricultural University*. Anand.

Appiani, M., Cattaneo, C., & Laureati, M. (2023). Sensory properties and consumer acceptance of plant-based meat, dairy, fish and eggs analogs: A systematic review. *Frontiers in Sustainable Food Systems, 7*, 1268068. DOI: 10.3389/fsufs.2023.1268068

Arab, M., Yousefi, M., Khanniri, E., Azari, M., Ghasemzadeh-Mohammadi, V., & Mollakhalili-Meybodi, N. (2023). A comprehensive review on yogurt syneresis: Effect of processing conditions and added additives. *Journal of Food Science and Technology, 60*(6), 1656–1665. DOI: 10.1007/s13197-022-05403-6 PMID: 37187980

Arrieta, A., Barrera, I., & Mendoza, J. (2023). Application of a Polypyrrole Sensor Array Integrated into a Smart Electronic Tongue for the Discrimination of Milk Adulterated with Sucrose. *International Journal of Technology, 14*(1), 90–99. DOI: 10.14716/ijtech.v14i1.5613

Arya, L., Sharma, Y. K., & Kumar, R. (2023). Towards a greener tomorrow: IoT-enabled smart environment monitoring systems. *2023 International Conference on Advances in Computation, Communication and Information Technology (ICAICCIT)*, 1112-1117. DOI: 10.1109/ICAICCIT60255.2023.10465894

Arya, L., Sharma, Y. K., Kumar, R., Padmanaban, H., Devi, S., & Tyagi, L. K. (2023). Maximizing IoT security: An examination of cryptographic algorithms. *2023 International Conference on Power, Energy, Environment & Intelligent Control (PEEIC)*, 1548-1552. DOI: 10.1109/PEEIC59336.2023.10451210

Asase, R. V., & Glukhareva, T. V. (2024). Production and application of xanthan gum—prospects in the dairy and plant-based milk food industry: A review. *Food Science and Biotechnology, 33*(4), 749–767. DOI: 10.1007/s10068-023-01442-7 PMID: 38371690

Atik, D. S., Öztürk, H. İ., & Akın, N. (2024). Perspectives on the yogurt rheology. *International Journal of Biological Macromolecules, 263*, 130428. DOI: 10.1016/j.ijbiomac.2024.130428 PMID: 38403217

Attaran, M. (2020). Digital technology enablers and their implications for supply chain management. *Supply Chain Forum: An International Journal, 21*(3), 158–172. DOI: 10.1080/16258312.2020.1751568

Aung, M. M., & Chang, Y. S. (2014). Traceability in a food supply chain: Safety and quality perspectives. *Food Control, 39*, 172–184. DOI: 10.1016/j.foodcont.2013.11.007

Avhale, G. V., Shaligram, A. D., & Gharpure, D. C. (2021). IoT based intelligent tracking, early-warning & management (ITEM) tool for efficient milk transportation in dairy industry. *AIP Conference Proceedings*, *2335*(1), 60003. DOI: 10.1063/5.0043687

Ayari, F., Mirzaee-Ghaleh, E., Rabbani, H., & Heidarbeigi, K. (2018). Detection of the adulteration in pure cow ghee by electronic nose method (case study: Sunflower oil and cow body fat). *International Journal of Food Properties*, *21*(1), 1670–1679. DOI: 10.1080/10942912.2018.1505755

Aydogdu, T., O'Mahony, J. A., & McCarthy, N. A. (2023). pH, the Fundamentals for Milk and Dairy Processing: A Review. *Dairy*, *4*(3), 395–409. DOI: 10.3390/dairy4030026

Azad, T., & Ahmed, S. (2016). Common milk adulteration and their detection techniques. *International Journal of Food Contamination*, *3*(1), 1–9. DOI: 10.1186/s40550-016-0045-3

Babiak, W., & Krzemińska, I. (2021). Extracellular polymeric substances (EPS) as microalgal bioproducts: A review of factors affecting EPS synthesis and application in flocculation processes. *Energies*, *14*(13), 4007. DOI: 10.3390/en14134007

Bahn, R. A., Yehya, A. A. K., & Zurayk, R. (2021). Digitalization for Sustainable Agri-Food Systems : Potential, Status, and Risks for the MENA Region. *Sustainability (Basel)*, *13*(6), 3223. DOI: 10.3390/su13063223

Baldin, M., Breunig, T., Cue, R., De Vries, A., Doornink, M., Drevenak, J., Fourdraine, R., George, R., Goodling, R., Greenfield, R., Jorgensen, M. W., Lenkaitis, A., Reinemann, D., Saha, A., Sankaraiah, C., Shahinfar, S., Siberski, C., Wade, K. M., Zhang, F., & Cabrera, V. E. (2021). Integrated Decision Support Systems (IDSS) for Dairy Farming: A Discussion on How to Improve Their Sustained Adoption. *Animals (Basel)*, *11*(7), 2025. Advance online publication. DOI: 10.3390/ani11072025 PMID: 34359153

Balivo, A., Cipolletta, S., Tudisco, R., Iommelli, P., Sacchi, R., & Genovese, A. (2023). Electronic nose analysis to detect milk obtained from pasture-raised goats. *Applied Sciences (Basel, Switzerland)*, *13*(2), 861. DOI: 10.3390/app13020861

Banhazi, T. M., Lewis, B., & Tscharke, M. (2011). The Development and Commercialisation Aspects of a Practical Feed Intake Measurement Instrumentation to be Used in Livestock Buildings. *Australian Journal of Multi-Disciplinary Engineering*, *8*(2), 131–138. DOI: 10.1080/14488388.2011.11464832

Banti, M. (2020). Food adulteration and some methods of detection [review]. *International Journal of Nutrition and Food Sciences*, *9*(3), 86–94. DOI: 10.11648/j.ijnfs.20200903.13

Basiri, S. (2021). Applications of microbial exopolysaccharides in the food Industry. *Avicenna Journal of Medical Biochemistry*, *9*(2), 107–120. DOI: 10.34172/ajmb.2021.16

Beber, C. L., Lakner, S., & Skevas, I. (2021). Organizational forms and technical efficiency of the dairy processing industry in Southern Brazil. *Agricultural and Food Economics*, *9*(1), 23. DOI: 10.1186/s40100-021-00195-3

Belhadi, A., Kamble, S., Fosso Wamba, S., & Queiroz, M. M. (2022). Building supply-chain resilience: An artificial intelligence-based technique and decision-making framework. *International Journal of Production Research*, *60*(14), 4487–4507. DOI: 10.1080/00207543.2021.1950935

Berthold-Pluta, A. M., Pluta, A. S., Garbowska, M., & Stasiak-Różańska, L. (2019). Exopolysaccharide-producing lactic acid bacteria–health-promoting properties and application in the dairy industry. *Postepy Mikrobiologii*, *58*(2), 191–204. DOI: 10.21307/PM-2019.58.2.191

Bewley, J. (2010). Precision Dairy Farming : Advanced Analysis Solutions for Future Profitability.

Bewley, J. (2010). Precision Dairy Farming : Advanced Analysis Solutions for Future Profitability. Academic Press.

Bhat, M. A., Rather, M. Y., Singh, P., Hassan, S., & Hussain, N. (2024). Advances in Smart Food Authentication for Enhanced Safety and Quality. *Trends in Food Science & Technology*, 104800.

Bianchi, M. C., Bava, L., Sandrucci, A., Tangorra, F. M., Tamburini, A., Gislon, G., & Zucali, M. (2022). Diffusion of precision livestock farming technologies in dairy cattle farms. *Animal*, *16*(11), 100650. DOI: 10.1016/j.animal.2022.100650 PMID: 36220000

Bihola, A., Sharma, H., Chaudhary, M., Bumbadiya, M., Kumar, D., & Adil, S. (2024). Recent developments in cheese technologies. *Food Reviews International*, 1–35.

Bosakova-Ardenska, A. (2024). Recent Trends in Computer Vision for Cheese Quality Evaluation. *Engineering Proceedings*, *60*(1), 12.

Botondi, R., Lembo, M., Carboni, C., & Eramo, V. (2023). The use of ozone technology: An eco–friendly method for the sanitization of the dairy supply chain. *Foods, 12*(5), 987. DOI: 10.3390/foods12050987 PMID: 36900504

Boukria, O., Boudalia, S., Bhat, Z. F., Hassoun, A., & Aït-Kaddour, A. (2023). Evaluation of the adulteration of camel milk by non-camel milk using multispectral image, fluorescence and infrared spectroscopy. *Spectrochimica Acta. Part A: Molecular and Biomolecular Spectroscopy, 300*, 122932. DOI: 10.1016/j.saa.2023.122932 PMID: 37270971

Brooks, C., Parr, L., Smith, J. M., Buchanan, D., Snioch, D., & Hebishy, E. (2021). A review of food fraud and food authenticity across the food supply chain, with an examination of the impact of the COVID-19 pandemic and Brexit on food industry. *Food Control, 130*, 108171. DOI: 10.1016/j.foodcont.2021.108171

Bryndin, E. (2020). Formation and Management of Industry 5.0 by Systems with Artificial Intelligence and Technological Singularity. *American Journal of Mechanical and Industrial Engineering*, *5*(2), 24–30. DOI: 10.11648/j.ajmie.20200502.12

Bueno, L., De Araujo, W. R., Salles, M. O., Kussuda, M. Y., & Paixão, T. R. (2014). Voltammetric electronic tongue for discrimination of milk adulterated with urea, formaldehyde and melamine. *Chemosensors (Basel, Switzerland)*, *2*(4), 251–266. DOI: 10.3390/chemosensors2040251

Bunce, B. (2020). Dairy Joint Ventures in South Africa's Land and Agrarian Reform Programme: Who Benefits? In *Land* (Vol. 9, Issue 9). DOI: 10.3390/land9090328

Burmistrov, D. E., Pavkin, D. Y., Khakimov, A. R., Ignatenko, D. N., Nikitin, E. A., Lednev, V. N., Lobachevsky, Y. P., Gudkov, S. V., & Zvyagin, A. V. (2021). Application of Optical Quality Control Technologies in the Dairy Industry: An Overview. In *Photonics* (Vol. 8, Issue 12). DOI: 10.3390/photonics8120551

Caballero-Agosto, E. R., Sierra-Vega, N. O., Rolon-Ocasio, Y., Hernandez-Rivera, S. P., Infante-Degró, R. A., Fontalvo-Gomez, M., Pacheco-Londoño, L. C., & Infante-Castillo, R. (2024). Detection and quantification of corn starch and wheat flour as adulterants in milk powder by near- and mid-infrared spectroscopy coupled with chemometric routines. *Food Chemistry Advances*, *4*, 100582. DOI: 10.1016/j.focha.2023.100582

Cabrera, V. E., & Fadul-Pacheco, L. (2021). Future of dairy farming from the Dairy Brain perspective: Data integration, analytics, and applications. *International Dairy Journal*, *121*, 105069. DOI: 10.1016/j.idairyj.2021.105069

Carrillo-Lopez, L. M., Garcia-Galicia, I. A., Tirado-Gallegos, J. M., Sanchez-Vega, R., Huerta-Jimenez, M., Ashokkumar, M., & Alarcon-Rojo, A. D. (2021). Recent advances in the application of ultrasound in dairy products: Effect on functional, physical, chemical, microbiological and sensory properties. *Ultrasonics Sonochemistry*, *73*, 105467. DOI: 10.1016/j.ultsonch.2021.105467 PMID: 33508590

Cedillos, R., Aleman, R. S., Page, R., Olson, D. W., Boeneke, C., Prinyawiwatkul, W., & Aryana, K. (2024). Influence of Hesperidin on the Physico-Chemical, Microbiological and Sensory Characteristics of Frozen Yogurt. *Foods*, *13*(5), 808. DOI: 10.3390/foods13050808 PMID: 38472921

Cen, H., & He, Y. (2007). Theory and application of near infrared reflectance spectroscopy in determination of food quality. *Trends in Food Science & Technology*, *18*(2), 72–83. DOI: 10.1016/j.tifs.2006.09.003

Chaisuwan, W., Jantanasakulwong, K., Wangtueai, S., Phimolsiripol, Y., Chaiyaso, T., Techapun, C., Phongthai, S., You, S. G., Regenstein, J. M., & Seesuriyachan, P. (2020). Microbial exopolysaccharides for immune enhancement: Fermentation, modifications and bioactivities. *Food Bioscience*, *35*, 100564. DOI: 10.1016/j.fbio.2020.100564

Chan, S. L., Kwok, T., Xu, N., Bo, T., & Huang, T. (2025). Imaged capillary isoelectric focusing and online mass spectrometry for milk whey protein characterization in dairy products. *Analytical Biochemistry*, *699*, 115765. DOI: 10.1016/j.ab.2025.115765 PMID: 39778730

Chaudhry, H. R., Muhammad, K., & Rabbani, M. (2015). Laboratory manual quality control of milk: quality control of milk. *Amazon2015 Aug*.

Chen, H., Tan, C., Lin, Z., & Wu, T. (2017, February 15). Detection of melamine adulteration in milk by near-infrared spectroscopy and one-class partial least squares. *Spectrochimica Acta. Part A: Molecular and Biomolecular Spectroscopy*, *173*, 832–836. DOI: 10.1016/j.saa.2016.10.051 PMID: 27816741

Chen, K., Magri, G., Grasset, E. K., & Cerutti, A. (2020). Rethinking mucosal antibody responses: IgM, IgG and IgD join IgA. *Nature Reviews. Immunology*, *20*(7), 427–441. DOI: 10.1038/s41577-019-0261-1 PMID: 32015473

Chen, R. K., Chang, L. W., Chung, Y. Y., Lee, M. H., & Ling, Y. C. (2004). Quantification of cow milk adulteration in goat milk using high-performance liquid chromatography with electrospray ionization mass spectrometry. *Rapid Communications in Mass Spectrometry*, *18*(10), 1167–1171. DOI: 10.1002/rcm.1460 PMID: 15150843

Chi, F., Tan, Z., Wang, Q., Yang, L., & Gu, X. (2024). Tibetan Butter and Indian Ghee: A Review on Their Production and Adulteration. *Agriculture*, *14*(9), 1533. DOI: 10.3390/agriculture14091533

Chi, Y., Dong, Y., Wang, Z. J., Yu, F. R., & Leung, V. C. M. (2022). Knowledge-Based Fault Diagnosis in Industrial Internet of Things: A Survey. *IEEE Internet of Things Journal*, *9*(15), 12886–12900. DOI: 10.1109/JIOT.2022.3163606

Choi, K., Yi, J., Park, C., & Yoon, S. (2021). Deep Learning for Anomaly Detection in Time-Series Data: Review, Analysis, and Guidelines. *IEEE Access : Practical Innovations, Open Solutions*, *9*, 120043–120065. DOI: 10.1109/ACCESS.2021.3107975

Chowdhury, M. A. H., Reem, C. S. A., Rahman, S. M., & Sarkar, F. (2025). Comprehensive approaches for ensuring microbial safety in the dairy industry: Monitoring systems, inhibitory strategies and future prospects. *Food Control*, *168*, 110894. DOI: 10.1016/j.foodcont.2024.110894

Cirak, O., Icyer, N. C., & Durak, M. Z. (2018). Rapid detection of adulteration of milks from different species using Fourier Transform Infrared Spectroscopy (FTIR). *The Journal of Dairy Research*, *85*(2), 222–225. DOI: 10.1017/S0022029918000201 PMID: 29785908

Cockburn, M. (2020). Review: Application and Prospective Discussion of Machine Learning for the Management of Dairy Farms. In *Animals* (Vol. 10, Issue 9). DOI: 10.3390/ani10091690

Dabbene, F., Gay, P., & Tortia, C. (2014). Traceability issues in food supply chains: A review. *Biosystems Engineering*, *120*, 65–80. DOI: 10.1016/j.biosystemseng.2013.09.006

Dahiya, P., Kumari, S., Behl, M., Kashyap, A., Kumari, D., Thakur, K., & Walia, A. (2024). Guardians of the Gut: Harnessing the Power of Probiotic Microbiota and Their Exopolysaccharides to Mitigate Heavy Metal Toxicity in Human for Better Health. *Probiotics and Antimicrobial Proteins*, *16*(6), 1–17. DOI: 10.1007/s12602-024-10281-9 PMID: 38733461

Danezis, G. P., Tsagkaris, A. S., Brusic, V., & Georgiou, C. A. (2016). Food authentication: State of the art and prospects. *Current Opinion in Food Science*, *10*, 22–31. DOI: 10.1016/j.cofs.2016.07.003

Das, N., & Hooda, A. (2023). Chemistry and Different Aspects of Ice Cream. In *The Chemistry of Milk and Milk Products* (pp. 65–86). Apple Academic Press. DOI: 10.1201/9781003340706-5

Demestichas, K., Peppes, N., & Alexakis, T. (2020). Survey on Security Threats in Agricultural IoT and Smart Farming. *Sensors (Basel), 20*(22), 6458. Advance online publication. DOI: 10.3390/s20226458 PMID: 33198160

Deng, Z., Wang, T., Zheng, Y., Zhang, W., & Yun, Y. H. (2024). Deep learning in food authenticity: Recent advances and future trends. *Trends in Food Science & Technology, 144*, 104344. DOI: 10.1016/j.tifs.2024.104344

Dimitri, C., & Nehring, R. (2022). Thirty years of organic dairy in the United States: The influences of farms, the market and the organic regulation. *Renewable Agriculture and Food Systems, 37*(6), 588–602. DOI: 10.1017/S1742170522000254

Dineva, K., & Atanasova, T. (2023). Health Status Classification for Cows Using Machine Learning and Data Management on AWS Cloud. *Animals (Basel), 13*(20), 3254. DOI: 10.3390/ani13203254 PMID: 37893978

Duan, K., Onyeaka, H., & Pang, G. (2024). Leveraging blockchain to tackle food fraud: Innovations and obstacles. *Journal of Agriculture and Food Research, 18*, 101429. DOI: 10.1016/j.jafr.2024.101429

Duboux, S., Van Wijchen, M., & Kleerebezem, M. (2021). The possible link between manufacturing and probiotic efficacy; a molecular point of view on Bifidobacterium. *Frontiers in Microbiology, 12*, 812536. DOI: 10.3389/fmicb.2021.812536 PMID: 35003044

Du, C., Zhao, X., Chu, C., Nan, L., Ren, X., Yan, L., Zhang, X. J., Zhang, S. J., & Teng, Z. (2025). Identification and quantification of goat milk adulteration using mid-infrared spectroscopy and chemometrics. *Spectrochimica Acta. Part A: Molecular and Biomolecular Spectroscopy, 324*, 124969. DOI: 10.1016/j.saa.2024.124969 PMID: 39153347

Du, L., Lu, W., Zhang, Y., Gao, B., & Yu, L. (2020). Detection of milk powder in liquid whole milk using hydrolyzed peptide and intact protein mass spectral fingerprints coupled with data fusion technologies. *Food Science & Nutrition, 8*(3), 1471–1479. DOI: 10.1002/fsn3.1430 PMID: 32180956

Džermeikaitė, K., Bačėninaitė, D., & Antanaitis, R. (2023). Innovations in Cattle Farming: Application of Innovative Technologies and Sensors in the Diagnosis of Diseases. *Animals (Basel), 13*(5), 780. Advance online publication. DOI: 10.3390/ani13050780 PMID: 36899637

Facure, M. H., Braunger, M. L., Mercante, L. A., Paterno, L. G., Riul, A., & Correa, D. S. (2022). Electrical Impedance-Based Electronic Tongues. *Encyclopedia of Sensors and Biosensors: Volume 1-4,* 567-590.

Fanelli, V., Mascio, I., Miazzi, M. M., Savoia, M. A., De Giovanni, C., & Montemurro, C. (2021). Molecular approaches to agri-food traceability and authentication: An updated review. *Foods*, *10*(07), 1644. DOI: 10.3390/foods10071644 PMID: 34359514

Feng, H., Wang, X., Duan, Y., Zhang, J., & Zhang, X. (2020). Applying blockchain technology to improve agri-food traceability: A review of development methods, benefits and challenges. *Journal of Cleaner Production*, *260*, 121031. DOI: 10.1016/j.jclepro.2020.121031

Fernández González, A., Badía Laíño, R., Costa-Fernández, J. M., & Soldado, A. (2024). Progress and Challenge of Sensors for Dairy Food Safety Monitoring. *Sensors (Basel)*, *24*(5), 1383. DOI: 10.3390/s24051383 PMID: 38474919

Fernando, I., Fei, J., Cahoon, S., & Close, D. C. (2024). A review of the emerging technologies and systems to mitigate food fraud in supply chains. *Critical Reviews in Food Science and Nutrition*, •••, 1–28. DOI: 10.1080/10408398.2024.2405840 PMID: 39356551

Ferreira, J. L. A., de Almeida, L. F., da Silva Simões, S., Diniz, P. H. G. D., & de Sousa Fernandes, D. D. (2025). Raman spectroscopy-based authentication of powder goat milk adulteration with cow milk. *Food Control*, *167*, 110800. DOI: 10.1016/j.foodcont.2024.110800

Ferreira, M. M., Marins-Gonçalves, L., & De Souza, D. (2024). An integrative review of analytical techniques used in food authentication: A detailed description for milk and dairy products. *Food Chemistry*, *457*, 140206. DOI: 10.1016/j.foodchem.2024.140206 PMID: 38936134

Fidan, I., Huseynov, O., Ali, M. A., Alkunte, S., Rajeshirke, M., Gupta, A., Hasanov, S., Tantawi, K., Yasa, E., Yilmaz, O., Loy, J., Popov, V., & Sharma, A. (2023). Recent Inventions in Additive Manufacturing: Holistic Review. In *Inventions* (Vol. 8, Issue 4). DOI: 10.3390/inventions8040103

Fiorino, G. M., Garino, C., Arlorio, M., Logrieco, A. F., Losito, I., & Monaci, L. (2018). Overview on untargeted methods to combat food frauds: A focus on fishery products. *Journal of Food Quality*, *2018*(1), 1581746. DOI: 10.1155/2018/1581746

Fisher, O. J., Watson, N. J., Escrig, J. E., Witt, R., Porcu, L., Bacon, D., Rigley, M., & Gomes, R. L. (2020). Considerations, challenges and opportunities when developing data-driven models for process manufacturing systems. *Computers & Chemical Engineering*, *140*, 106881. DOI: 10.1016/j.compchemeng.2020.106881

Fitcows. (n.d.). Milk monitoring system using IoT to check milk's quality. Retrieved from https://www.fitcows.com/blogs/milk-monitoring-system-using-iot/

Food and Agriculture Organization (FAO). (2023). Global milk production statistics. Retrieved from https://www.fao.org/dairy-production-products/production/milk-production/en

Frigerio, J., Campone, L., Giustra, M. D., Buzzelli, M., Piccoli, F., Galimberti, A., Cannavacciuolo, C., Ouled Larbi, M., Colombo, M., Ciocca, G., & Labra, M. (2024). Convergent technologies to tackle challenges of modern food authentication. *Heliyon, 10*(11), e32297. DOI: 10.1016/j.heliyon.2024.e32297 PMID: 38947432

Galukande, E., Mulindwa, H., Wurzinger, M., Roschinsky, R., Mwai, A. O., & Sölkner, J. (2013). Cross-breeding cattle for milk production in the tropics: Achievements, challenges, and opportunities. *Animal Genetic Resources, 52*, 111–125. DOI: 10.1017/S2078633612000471

Gehlot, A., Malik, P. K., Singh, R., Akram, S. V., & Alsuwian, T. (2022). Dairy 4.0: Intelligent Communication Ecosystem for the Cattle Animal Welfare with Blockchain and IoT Enabled Technologies. *Applied Sciences (Basel, Switzerland), 12*(14), 7316. DOI: 10.3390/app12147316

Germani, L., Mecarelli, V., Baruffa, G., Rugini, L., & Frescura, F. (2019). An IoT architecture for continuous livestock monitoring using LoRa LPWAN. *Electronics (Basel), 8*(12), 1435. DOI: 10.3390/electronics8121435

Ghosal, A., Sagong, S., Halder, S., Sahabandu, K., Conti, M., Poovendran, R., & Bushnell, L. (2020). Truck platoon security: State-of-the-art and road ahead. *Computer Networks, 185*, 107658. DOI: 10.1016/j.comnet.2020.107658

Global Dairy Industry and Trends - Feed & Additive Magazine. (n.d.). Retrieved November 26, 2024, from https://www.feedandadditive.com/global-dairy-industry-and-trends/

Goh, K. K., Teo, A., Sarkar, A., & Singh, H. (2020). Milk protein-polysaccharide interactions. In *Milk proteins* (pp. 499–535). Elsevier. DOI: 10.1016/B978-0-12-815251-5.00013-X

Gowda, D. D., Sharma, A., Suneel, S., Kumar, N. A., Sharma, A., Suneel, S., & Manikandan, V. (2024). *Design and Implementation of an IoT-Based Control System for Precision Food Manufacturing*. DOI: 10.1007/978-981-97-5146-4_29

Goyal, K., Kumar, P., & Verma, K. (2022). Food adulteration detection using artificial intelligence: A systematic review. *Archives of Computational Methods in Engineering, 29*(1), 397–426. DOI: 10.1007/s11831-021-09600-y

Groumpos, P. P. (2021). A Critical Historical and Scientific Overview of all Industrial Revolutions. *IFAC-PapersOnLine*, *54*(13), 464–471. DOI: 10.1016/j.ifacol.2021.10.492 PMID: 38620687

Gunturi, M., Kotha, H. D., & Srinivasa Reddy, M. (2018). An overview of internet of things. *Journal of Advanced Research in Dynamical and Control Systems*, *10*(9), 659–665.

Gupta, B. B., & Quamara, M. (2020). An overview of Internet of Things (IoT): Architectural aspects, challenges, and protocols. *Concurrency and Computation*, *32*(21), e4946. DOI: 10.1002/cpe.4946

Gupta, P., & Roy, S. (2023). Enhancing dairy productivity with IoT. *Advances in IoT Systems*, *5*(4), 299–315.

Gu, Z., Meng, S., Wang, Y., Lyu, B., Li, P., & Shang, N. (2023). A novel bioactive postbiotics: From microbiota-derived extracellular nanoparticles to health promoting. *Critical Reviews in Food Science and Nutrition*, *63*(24), 6885–6899. DOI: 10.1080/10408398.2022.2039897 PMID: 35179102

Hadi, M. S., Sugiono, B. S. R., Mizar, M. A., Witjoro, A., & Irvan, M. (2024). Enhancing low-temperature long-time milk pasteurization process with a C4.5 algorithm-driven AIoT system for real-time decision-making. *Journal of Food Process Engineering*, *47*(4), e14606. DOI: 10.1111/jfpe.14606

Halachmi, I., Guarino, M., Bewley, J., & Pastell, M. (2019). Smart Animal Agriculture : Application of Real-Time Sensors to Improve Animal Well-Being and Production. *Annual Review of Animal Biosciences*, *7*(1), 403–425. DOI: 10.1146/annurev-animal-020518-114851 PMID: 30485756

Haldar, L., Raghu, H. V., & Ray, P. R. (2022). Milk and milk product safety and quality assurance for achieving better public health outcomes. *Agriculture. Livestock Production and Aquaculture: Advances for Smallholder Farming Systems*, *1*, 217–259. DOI: 10.1007/978-3-030-93258-9_13

Han, H.-S., Shin, J.-S., Song, Y.-R., Rhee, Y. K., Cho, C.-W., Ryu, J. H., Inn, K.-S., Hong, H.-D., & Lee, K.-T. (2020). Immunostimulatory effects of polysaccharides isolated from young barley leaves (Hordeum vulgare L.) with dual activation of Th1 and Th2 in splenic T cells and cyclophosphamide-induced immunosuppressed mice. *International Journal of Biological Macromolecules*, *147*, 954–964. DOI: 10.1016/j.ijbiomac.2019.10.062 PMID: 31730956

Han, H., Wang, Z., Li, C., Ma, Z., Yang, Z., & Ma, X. (2019). Purity detection of goat milk based on electronic tongue and improved artificial fish swarm optimized extreme learning machine. *IFAC-PapersOnLine*, *52*(30), 391–396. DOI: 10.1016/j.ifacol.2019.12.588

Harini, V. K., Meher, S. R., & Alex, Z. C. (2024). A novel refractive index based-fiber optic sensor for milk adulteration detection. *Optical Materials*, *154*, 115810. DOI: 10.1016/j.optmat.2024.115810

Hassoun, A., Dankar, I., Bhat, Z., & Bouzembrak, Y. (2024). Unveiling the relationship between food unit operations and food industry 4.0: A short review. *Heliyon*, *10*(20), e39388. DOI: 10.1016/j.heliyon.2024.e39388 PMID: 39492883

Haug, A., Høstmark, A. T., & Harstad, O. M. (2007). Bovine milk in human nutrition–a review. *Lipids in Health and Disease*, *6*(1), 1–16. DOI: 10.1186/1476-511X-6-25 PMID: 17894873

Hazra, A., Adhikari, M., Amgoth, T., & Srirama, S. (2021). A Comprehensive Survey on Interoperability for IIoT: Taxonomy, Standards, and Future Directions. *ACM Computing Surveys*, *55*(1), 35. DOI: 10.1145/3485130

Heema, R., .S, S. & Gnanalakshmi, K. S. (2022). An Insight in to the Automation of the Dairy Industry: A Review. *Asian Journal of Dairy and Food Research*. DOI: 10.18805/ajdfr.DR-1856

He, H., Sun, D. W., Pu, H., Chen, L., & Lin, L. (2019). Applications of Raman spectroscopic techniques for quality and safety evaluation of milk: A review of recent developments. *Critical Reviews in Food Science and Nutrition*, *59*(5), 770–793. DOI: 10.1080/10408398.2018.1528436 PMID: 30614242

Helo, P., & Hao, Y. (2021). Artificial intelligence in operations management and supply chain management: An exploratory case study. *Production Planning and Control*, *33*(16), 1–18. DOI: 10.1080/09537287.2021.1882690

Hepfer, M., & Lawrence, T. (2022). The Heterogeneity of Organizational Resilience: Exploring functional, operational and strategic resilience. *Organization Theory*, *3*(1). DOI: 10.1177/26317877221074701

He, Y., Bai, X., Xiao, Q., Liu, F., Zhou, L., & Zhang, C. (2021). Detection of adulteration in food based on nondestructive analysis techniques: A review. *Critical Reviews in Food Science and Nutrition*, *61*(14), 2351–2371. DOI: 10.1080/10408398.2020.1777526 PMID: 32543218

Hossen, R., Hossain, M. S., Ahmed, S., Rahman, M. M., & Sen, S. (2024). Illegal food chemicals sensing with photonic crystal fiber sensor in the terahertz spectrum. *Sensing and Bio-Sensing Research*, *44*, 100643. DOI: 10.1016/j.sbsr.2024.100643

Hostiou, N., Allain, C., Chauvat, S., Turlot, A., Pineau, C., & Fagon, J. (2014). L'élevage de précision : Quelles conséquences pour le travail des éleveurs? *INRA Productions Animales*, *27*(2), 113–122. DOI: 10.20870/productions-animales.2014.27.2.3059

Hostiou, N., Fagon, J., Chauvat, S., Turlot, A., Kling-Eveillard, F., Boivin, X., & Allain, C. (2017). Impact of precision livestock farming on work and human- animal interactions on dairy farms. A review. *Biotechnologie, Agronomie, Société et Environnement*, 268–275. Advance online publication. DOI: 10.25518/1780-4507.13706

Hu, T., Zhang, J., Zhang, X., Chen, Y., Zhang, R., & Guo, K. (2023). The development of smart dairy farm system and its application in nutritional grouping and mastitis prediction. *Animals (Basel)*, *13*(5), 804. DOI: 10.3390/ani13050804 PMID: 36899660

Iftekhar, A., & Cui, X. (2021). Blockchain-Based Traceability System That Ensures Food Safety Measures to Protect Consumer Safety and COVID-19 Free Supply Chains. In *Foods* (Vol. 10, Issue 6). DOI: 10.3390/foods10061289

Iftekhar, A., Cui, X., Hassan, M., & Afzal, W. (2020). Application of Blockchain and Internet of Things to Ensure Tamper-Proof Data Availability for Food Safety. *Journal of Food Quality*, *2020*(1), 5385207. DOI: 10.1155/2020/5385207

Imam, M., & Nagpal, K. (2023). The electronic tongue: An advanced taste-sensing multichannel sensory tool with global selectivity for application in the pharmaceutical and food industry. *Pharmaceutical Development and Technology*, *28*(3-4), 318–332. DOI: 10.1080/10837450.2023.2194989 PMID: 36987792

Ionescu, A. D., Cîrîc, A. I., & Begea, M. (2023). A review of milk frauds and adulterations from a technological perspective. *Applied Sciences (Basel, Switzerland)*, *13*(17), 9821. DOI: 10.3390/app13179821

IoT ONE. (n.d.). IoT-based milk production application. Retrieved from https://www.iotone.com/case-study/iot-based-milk-production-application/c602

IoT ONE. (n.d.). IoT-based milk production application. Retrieved from https://www.iotone.com/case-study/iot-based-milk-production-application/c602Foo

Ivanov, D., Tang, C. S., Dolgui, A., Battini, D., & Das, A. (2021). Researchers' perspectives on Industry 4.0: Multi-disciplinary analysis and opportunities for operations management. *International Journal of Production Research*, *59*(7), 2055–2078. DOI: 10.1080/00207543.2020.1798035

Jadhav, A. R., R., S. K. M. P. & Pachamuthu, R. (. (2021). Development of a Novel IoT-Enabled Power- Monitoring Architecture With Real-Time Data Visualization for Use in Domestic and Industrial Scenarios. *IEEE Transactions on Instrumentation and Measurement*, *70*, 1–14. DOI: 10.1109/TIM.2020.3028437

Jaseera, K., & Abdulla, T. (2021). Microbial EPS as Immunomodulatory Agents. *Microbial Exopolysaccharides as Novel and Significant Biomaterials*, 235-264.

Jerhamre, E., Carlberg, C. J. C., & van Zoest, V. (2022). Exploring the susceptibility of smart farming: Identified opportunities and challenges. *Smart Agricultural Technology*, *2*, 100026. DOI: 10.1016/j.atech.2021.100026

Jiang, B., Wei, M., He, Y., Wang, Y., Zhang, Q., Wei, H., & Tao, X. (2024). Antitumor effect of exopolysaccharide from Lactiplantibacillus plantarum WLPL09 on melanoma mice via regulating immunity and gut microbiota. *International Journal of Biological Macromolecules*, *254*, 127624. DOI: 10.1016/j.ijbiomac.2023.127624 PMID: 37918610

Jin, H., Dong, G. M., Wu, H. Y., Yang, Y. R., Huang, M. Y., Wang, M. Y., & Yang, R. J. (2023). Identification of adulterated milk based on auto-correlation spectra. *Spectrochimica Acta. Part A: Molecular and Biomolecular Spectroscopy*, *286*, 121987. DOI: 10.1016/j.saa.2022.121987 PMID: 36265304

Jitsuwan, P., Klinsophon, N., Thanakiatkrai, P., & Kitpipit, T. (2024). Simple closed-system combining direct PCR and fluorescence for all types of milk and dairy product authentication. *Journal of Food Composition and Analysis*, *130*, 106178. DOI: 10.1016/j.jfca.2024.106178

Ji, Z., Zhang, J., Deng, C., Guo, T., Han, R., Yang, Y., Zang, C., & Chen, Y. (2024). Identification of pasteurized mare milk and powder adulteration with bovine milk using quantitative proteomics and metabolomics approaches. *Food Chemistry: X*, *22*, 101265. DOI: 10.1016/j.fochx.2024.101265 PMID: 38468636

Jurášková, D., Ribeiro, S. C., & Silva, C. C. (2022). Exopolysaccharides produced by lactic acid bacteria: From biosynthesis to health-promoting properties. *Foods*, *11*(2), 156. DOI: 10.3390/foods11020156 PMID: 35053888

Kalamaras, S. D., Tsitsimpikou, M.-A., Tzenos, C. A., Lithourgidis, A. A., Pitsikoglou, D. S., & Kotsopoulos, T. A. (2024). A Low-Cost IoT System Based on the ESP32 Microcontroller for Efficient Monitoring of a Pilot Anaerobic Biogas Reactor. *Applied Sciences (Basel, Switzerland)*, *15*(1), 34. DOI: 10.3390/app15010034

Kamboj, U., Kaushal, N., Mishra, S., & Munjal, N. (2020). Application of selective near infrared spectroscopy for qualitative and quantitative prediction of water adulteration in milk. *Materials Today: Proceedings*, *24*, 2449–2456. DOI: 10.1016/j.matpr.2020.03.775

Karouani, Y., & Elgarej, M. (2021). Milk-Run Collection Monitoring System Using the Internet of Things Based on Swarm Intelligence. *International Journal of Information Systems and Supply Chain Management*, *15*(3), 1–17. DOI: 10.4018/IJISSCM.290018

Karunathilaka, S. R., Yakes, B. J., He, K., Brückner, L., & Mossoba, M. M. (2018). First use of handheld Raman spectroscopic devices and on-board chemometric analysis for the detection of milk powder adulteration. *Food Control*, *92*, 137–146. DOI: 10.1016/j.foodcont.2018.04.046

Kavitha, D., & Ravikumar, S. (2021). IOT and context-aware learning-based optimal neural network model for real-time health monitoring. *Transactions on Emerging Telecommunications Technologies*, *32*(1), e4132. DOI: 10.1002/ett.4132

Khanna, A., Jain, S., Burgio, A., Bolshev, V., & Panchenko, V. (2022). Blockchain-Enabled Supply Chain platform for Indian Dairy Industry: Safety and Traceability. *Foods*, *11*(17), 2716. DOI: 10.3390/foods11172716 PMID: 36076901

Kharajinezhadian, R., Chaichi, M. J., Nazari, O., Lakouraj, M. M., & Hasantabar, V. (2023). Fraud monitoring using a new disposable photoluminescence sensor in milk. *Microchemical Journal*, *189*, 108437. DOI: 10.1016/j.microc.2023.108437

Khosroshahi, M. E., Patel, Y., & Woll-Morison, V. (2022). Non-destructive assessment of milk quality using pulsed UV photoacoustic, fluorescence and near FTIR spectroscopy. *Laser Physics Letters*, *19*(7), 075602. DOI: 10.1088/1612-202X/ac6fc5

Kjønås, K., & Wangen, G. (2024). In Fritsch, L., Hassan, I., & Paintsil, E. (Eds.), *Cybersecurity Challenges and Smart Technology Adoption in Norwegian Livestock Farming BT - Secure IT Systems* (pp. 107–125). Springer Nature Switzerland.

Konstantinidis, F. K., Balaska, V., Symeonidis, S., Psarommatis, F., Psomoulis, A., Giakos, G., Mouroutsos, S. G., & Gasteratos, A. (2023). Achieving Zero Defected Products in Diary 4.0 using Digital Twin and Machine Vision. *Proceedings of the 16th International Conference on PErvasive Technologies Related to Assistive Environments*, 528–534. DOI: 10.1145/3594806.3596554

Korcz, E., & Varga, L. (2021). Exopolysaccharides from lactic acid bacteria: Techno-functional application in the food industry. *Trends in Food Science & Technology*, *110*, 375–384. DOI: 10.1016/j.tifs.2021.02.014

Kozina, A. M., & Semkiv, L. P. (2020). Sustainable development of dairy farming through the use of digital technologies. *IOP Conference Series. Earth and Environmental Science, 613*(1), 012061. DOI: 10.1088/1755-1315/613/1/012061

Kreuzberger, D., Kühl, N., & Hirschl, S. (2023). Machine Learning Operations (MLOps): Overview, Definition, and Architecture. *IEEE Access : Practical Innovations, Open Solutions, 1*, 31866–31879. Advance online publication. DOI: 10.1109/ACCESS.2023.3262138

Kumar, A., & Joshi, R. (2021). IoT in dairy: A sustainable approach. *International Journal of Dairy Technology, 74*(3), 487–495.

Kumar, R. R., Rahman, F., Bora, B., & Shameeh, M. (2023). Importance and nutritive value of animal proteins in human diet. In *Processing Technologies and Food Protein Digestion* (pp. 1–25). Academic Press. DOI: 10.1016/B978-0-323-95052-7.00007-8

Kumar, S., Kumar, S., Bumbadia, M., & Singh, S. K. (2022). Prospects and functionality of bacterial exopolysaccharides in dairy foods: A Review: Bacterial exopolysaccharides in dairy. *Journal of AgriSearch, 9*(1), 1–5. DOI: 10.21921/jas.v9i01.9884

Kyaw, K. S., Adegoke, S. C., Ajani, C. K., Nwabor, O. F., & Onyeaka, H. (2024). Toward in-process technology-aided automation for enhanced microbial food safety and quality assurance in milk and beverages processing. *Critical Reviews in Food Science and Nutrition, 64*(6), 1715–1735. DOI: 10.1080/10408398.2022.2118660 PMID: 36066463

Lal, P. P., Prakash, A. A., Chand, A. A., Prasad, K. A., Mehta, U., Assaf, M. H., Mani, F. S., & Mamun, K. A. (2022). IoT integrated fuzzy classification analysis for detecting adulterants in cow milk. *Sensing and Bio-Sensing Research, 36*, 100486. DOI: 10.1016/j.sbsr.2022.100486

Leal, F., Chis, A. E., Caton, S., González–Vélez, H., García–Gómez, J. M., Durá, M., Sánchez–García, A., Sáez, C., Karageorgos, A., Gerogiannis, V. C., Xenakis, A., Lallas, E., Ntounas, T., Vasileiou, E., Mountzouris, G., Otti, B., Pucci, P., Papini, R., Cerrai, D., & Mier, M. (2021). Smart Pharmaceutical Manufacturing: Ensuring End-to-End Traceability and Data Integrity in Medicine Production. *Big Data Research, 24*, 100172. https://doi.org/https://doi.org/10.1016/j.bdr.2020.100172

Leal, M. R. S., Albuquerque, P. B. S., Rodrigues, N. E. R., dos Santos Silva, P. M., de Oliveira, W. F., dos Santos Correia, M. T., & Coelho, L. C. B. B. (2024). A review on the use of polysaccharides as thickeners in yogurts. *Carbohydrate Polymer Technologies and Applications, 8*, 100547. DOI: 10.1016/j.carpta.2024.100547

Ledo, J., Hettinga, K. A., & Luning, P. A. (2020). A customized assessment tool to differentiate safety and hygiene control practices in emerging dairy chains. *Food Control, 111*, 107072. DOI: 10.1016/j.foodcont.2019.107072

Lee, E., Seo, Y.-D., Oh, S.-R., & Kim, Y.-G. (2021). A Survey on Standards for Interoperability and Security in the Internet of Things. *IEEE Communications Surveys and Tutorials*, *23*(2), 1020–1047. DOI: 10.1109/COMST.2021.3067354

Lee, I., & Lee, K. (2015). The Internet of Things (IoT): Applications, investments, and challenges for enterprises. *Business Horizons*, *58*(4), 431–440. DOI: 10.1016/j.bushor.2015.03.008

Lekki-Jóźwiak, J., & Bąska, P. (2023). The Roles of Various Immune Cell Populations in Immune Response against Helminths. *International Journal of Molecular Sciences*, *25*(1), 420. DOI: 10.3390/ijms25010420 PMID: 38203591

Lelis, C. A., Galvan, D., Tessaro, L., de Andrade, J. C., Mutz, Y. S., & Conte-Junior, C. A. (2022). Fluorescence spectroscopy in tandem with chemometric tools applied to milk quality control. *Journal of Food Composition and Analysis*, *109*, 104515. DOI: 10.1016/j.jfca.2022.104515

Libelium. (n.d.). How a dairy farm increased their milk production 18% with IoT and machine learning. Retrieved from https://www.libelium.com/libeliumworld/success-stories/how-a-dairy-farm-increased-their-milk-production-18-with-iot-and-machine-learning/

Li, L., Rong, S., Wang, R., & Yu, S. (2021). Recent advances in artificial intelligence and machine learning for nonlinear relationship analysis and process control in drinking water treatment: A review. *Chemical Engineering Journal*, *405*, 126673. DOI: 10.1016/j.cej.2020.126673

Li, Y., Jin, L., & Chen, T. (2020). The effects of secretory IgA in the mucosal immune system. *BioMed Research International*, *2020*(1), 2032057. DOI: 10.1155/2020/2032057 PMID: 31998782

Lohmer, J., & Lasch, R. (2020). Blockchain in operations management and manufacturing: Potential and barriers. *Computers & Industrial Engineering*, *149*, 106789. DOI: 10.1016/j.cie.2020.106789

Lu, X., Chen, J., Guo, Z., Zheng, Y., Rea, M. C., Su, H., Zheng, X., Zheng, B., & Miao, S. (2019). Using polysaccharides for the enhancement of functionality of foods: A review. *Trends in Food Science & Technology*, *86*, 311–327. DOI: 10.1016/j.tifs.2019.02.024

Mabood, F., Ali, L., Boque, R., Abbas, G., Jabeen, F., Haq, Q. M. I., Hussain, J., Hamaed, A. M., Naureen, Z., Al-Nabhani, M., Khan, M. Z., Khan, A., & Al-Harrasi, A. (2020). Robust Fourier transformed infrared spectroscopy coupled with multivariate methods for detection and quantification of urea adulteration in fresh milk samples. *Food Science & Nutrition*, *8*(10), 5249–5258. DOI: 10.1002/fsn3.987 PMID: 33133527

Mahalik, N. P., & Nambiar, A. N. (2010). Trends in food packaging and manufacturing systems and technology. *Trends in Food Science & Technology*, *21*(3), 117–128. DOI: 10.1016/j.tifs.2009.12.006

Mahanti, N. K., Shivashankar, S., Chhetri, K. B., Kumar, A., Rao, B. B., Aravind, J., & Swami, D. V. (2024). Enhancing Food Authentication through E-Nose and E-Tongue Technologies: Current Trends and Future Directions. *Trends in Food Science & Technology*, *150*, 104574. DOI: 10.1016/j.tifs.2024.104574

Majcher, M. A., Kaczmarek, A., Klensporf-Pawlik, D., Pikul, J., & Jeleń, H. H. (2015). SPME-MS-based electronic nose as a tool for determination of authenticity of PDO cheese, Oscypek. *Food Analytical Methods*, *8*(9), 2211–2217. DOI: 10.1007/s12161-015-0114-x

Malik, M., Gahlawat, V. K., Mor, R. S., & Singh, M. K. (2024). Unlocking dairy traceability: Current trends, applications, and future opportunities. *Future Foods : a Dedicated Journal for Sustainability in Food Science*, *10*, 100426. DOI: 10.1016/j.fufo.2024.100426

Manivannan, D., Moni, S. S., & Zeadally, S. (2020). Secure authentication and privacy-preserving techniques in Vehicular Ad-hoc NETworks (VANETs). *Vehicular Communications*, *25*, 100247. DOI: 10.1016/j.vehcom.2020.100247

Marchi, B., Galati, C., & Zanoni, S. (2024). Management of spare parts for efficient maintenance: A case study in the dairy sector. *IFAC-PapersOnLine*, *58*(8), 389–394. DOI: 10.1016/j.ifacol.2024.08.152

Maroc Lait. (2022). maroclait. https://maroclait.ma/fr/stat.html

Mascini, M., & Tombelli, S. (2008). Biosensors for biomarkers in medical diagnostics. *Biomarkers*, *13*(7-8), 637–657. DOI: 10.1080/13547500802645905 PMID: 19061054

Mehany, T., Siddiqui, S. A., Olawoye, B., Olabisi Popoola, O., Hassoun, A., Manzoor, M. F., & Punia Bangar, S. (2024). Recent innovations and emerging technological advances used to improve quality and process of plant-based milk analogs. *Critical Reviews in Food Science and Nutrition*, *64*(20), 7237–7267. DOI: 10.1080/10408398.2023.2183381 PMID: 36861223

Mhapsekar, R. U., O'Shea, N., Davy, S., Kilbane, D., & Abraham, L. (2024). An edge-centric industrial IoT solution for smart dairy processing. *IEEE Internet of Things Magazine*, *7*(5), 80–87. DOI: 10.1109/IOTM.001.2400010

Michie, C., Andonovic, I., Davison, C., Hamilton, A., Tachtatzis, C., Jonsson, N., Duthie, C.-A., Bowen, J., & Gilroy, M. (2020). The Internet of Things enhancing animal welfare and farm operational efficiency. *The Journal of Dairy Research*, *87*(S1), 20–27. DOI: 10.1017/S0022029920000680 PMID: 33213573

Misiuk, M., & Zakhodym, M. (2023). Digitization as a tool for revitalizing the livestock industry. *Ekonomika APK*, *30*(4), 10–24. DOI: 10.32317/2221-1055.202304010

Moghaddam, H. N., Tamiji, Z., Lakeh, M. A., Khoshayand, M. R., & Mahmoodi, M. H. (2022). Multivariate analysis of food fraud: A review of NIR based instruments in tandem with chemometrics. *Journal of Food Composition and Analysis*, *107*, 104343. DOI: 10.1016/j.jfca.2021.104343

Mohammed, A., Potdar, V., & Quaddus, M. (2023). Exploring factors and impact of blockchain technology in the food supply chains: An exploratory study. *Foods*, *12*(10), 2052. DOI: 10.3390/foods12102052 PMID: 37238870

Morais, R., Mendes, J., Silva, R., Silva, N., Sousa, J. J., & Peres, E. (2021). A Versatile, Low-Power and Low-Cost IoT Device for Field Data Gathering in Precision Agriculture Practices. In *Agriculture* (Vol. 11, Issue 7). DOI: 10.3390/agriculture11070619

Mostafapour, S., Gharaghani, F. M., & Hemmateenejad, B. (2021). Converting electronic nose into opto-electronic nose by mixing MoS2 quantum dots with organic reagents: Application to recognition of aldehydes and ketones and determination of formaldehyde in milk. *Analytica Chimica Acta*, *1170*, 338654. DOI: 10.1016/j.aca.2021.338654 PMID: 34090585

Moszyk, K., Deja, M., & Dobrzyński, M. (2021). Automation of the Road Gate Operations Process at the Container Terminal—A Case Study of DCT Gdańsk SA. *Sustainability (Basel)*, *13*(11), 6291. DOI: 10.3390/su13116291

Moudgil, V., Hewage, K., Hussain, S. A., & Sadiq, R. (2023). Integration of IoT in building energy infrastructure: A critical review on challenges and solutions. *Renewable & Sustainable Energy Reviews*, *174*, 113121. DOI: 10.1016/j.rser.2022.113121

Moyer, D. C., DeVries, J. W., & Spink, J. (2017). The economics of a food fraud incident–Case studies and examples including Melamine in Wheat Gluten. *Food Control*, *71*, 358–364. DOI: 10.1016/j.foodcont.2016.07.015

Mrabet, H., Belguith, S., Alhomoud, A., & Jemai, A. (2020). A Survey of IoT Security Based on a Layered Architecture of Sensing and Data Analysis. In *Sensors* (Vol. 20, Issue 13). DOI: 10.3390/s20133625

Mu, F., Gu, Y., Zhang, J., & Zhang, L. (2020). Milk source identification and milk quality estimation using an electronic nose and machine learning techniques. *Sensors (Basel)*, *20*(15), 4238. DOI: 10.3390/s20154238 PMID: 32751425

Mullins, I. L., Truman, C. M., Campler, M. R., Bewley, J. M., & Costa, J. H. C. (2019). Validation of a Commercial Automated Body Condition Scoring System on a Commercial Dairy Farm. *Animals (Basel)*, *9*(6), 287. DOI: 10.3390/ani9060287 PMID: 31146374

Muninathan, C., Guruchandran, S., Viswanath Kalyan, A. J., & Ganesan, N. D. (2022). Microbial exopolysaccharides: Role in functional food engineering and gut-health management. *International Journal of Food Science & Technology*, *57*(1), 27–34. DOI: 10.1111/ijfs.15334

Murphy, S. I., Reichler, S. J., Martin, N. H., Boor, K. J., & Wiedmann, M. (2021). Machine Learning and Advanced Statistical Modeling Can Identify Key Quality Management Practices That Affect Postpasteurization Contamination of Fluid Milk. *Journal of Food Protection*, *84*(9), 1496–1511. DOI: 10.4315/JFP-20-431 PMID: 33770185

Nagaraj, S., Kathole, A. B., Arya, L., Tyagi, N., Goyal, S. B., Rajawat, A. S., Raboaca, M. S., Mihaltan, T. C., Verma, C., & Suciu, G. (2022). Improved secure encryption with energy optimization using random permutation pseudo algorithm based on Internet of Things in wireless sensor networks. *Energies*, *16*(8), 1–16. DOI: 10.3390/en16010008

Nagraik, R., Sharma, A., Kumar, D., Chawla, P., & Kumar, A. P. (2021). Milk adulterant detection: Conventional and biosensor based approaches: A review. *Sensing and Bio-Sensing Research*, *33*, 100433. DOI: 10.1016/j.sbsr.2021.100433

Najafi, M., Kaushal, V., & Visser, J. (2024). Operational Planning and Design Considerations for Underground Logistics Transportation in Texas. In *Infrastructures* (Vol. 9, Issue 8). DOI: 10.3390/infrastructures9080130

Naqvi, R., Soomro, T., Alzoubi, H., Ghazal, T. & Alshurideh, M. (2021). *The Nexus Between Big Data and Decision-Making: A Study of Big Data Techniques and Technologies*. DOI: 10.1007/978-3-030-76346-6_73

Nascimento, C. F., Santos, P. M., Pereira-Filho, E. R., & Rocha, F. R. (2017). Recent advances on determination of milk adulterants. *Food Chemistry, 221*, 1232–1244. DOI: 10.1016/j.foodchem.2016.11.034 PMID: 27979084

Neculai-Valeanu, A.-S., & Ariton, A.-M. (2022). Udder Health Monitoring for Prevention of Bovine Mastitis and Improvement of Milk Quality. *Bioengineering (Basel, Switzerland), 9*(11), 608. Advance online publication. DOI: 10.3390/bioengineering9110608 PMID: 36354519

Neethirajan, S. (2023). Artificial Intelligence and Sensor Technologies in Dairy Livestock Export: Charting a Digital Transformation. In *Sensors* (Vol. 23, Issue 16). DOI: 10.3390/s23167045

Neethirajan, S., & Kemp, B. (2021). Digital Livestock Farming. *Sensing and Bio-Sensing Research, 32*, 100408. DOI: 10.1016/j.sbsr.2021.100408

Nestlé Sustainability Reports. (2021). Water and energy optimization in dairy processing facilities. Retrieved from https://www.nestle.com

Newton, J. E., Nettle, R., & Pryce, J. E. (2020). Farming smarter with big data: Insights from the case of Australia's national dairy herd milk recording scheme. *Agricultural Systems, 181*, 102811. DOI: 10.1016/j.agsy.2020.102811

Nguyen, P.-T., Nguyen, T.-T., Bui, D.-C., Hong, P.-T., Hoang, Q.-K., & Nguyen, H.-T. (2020). Exopolysaccharide production by lactic acid bacteria: The manipulation of environmental stresses for industrial applications. *AIMS Microbiology, 6*(4), 451–469. DOI: 10.3934/microbiol.2020027 PMID: 33364538

Nikander, J., Manninen, O., & Laajalahti, M. (2020). Requirements for cybersecurity in agricultural communication networks. *Computers and Electronics in Agriculture, 179*, 105776. DOI: 10.1016/j.compag.2020.105776

Nižetić, S., Šolić, P., López-de-Ipiña González-de-Artaza, D., & Patrono, L. (2020). Internet of Things (IoT): Opportunities, issues and challenges towards a smart and sustainable future. *Journal of Cleaner Production, 274*, 122877. DOI: 10.1016/j.jclepro.2020.122877 PMID: 32834567

Novaes, A. G. N., Lima, O. F., De Cursi, J. E. S., Arias, J. A. C., & Santos, J. B. S. (2020). In Kotzab, H., & Pannek, J. (Eds.), *Predictive Manufacturing Tardiness Inference in OEM Milk-Run Operations BT - Dynamics in Logistics* (pp. 254–262). Springer International Publishing.

Nowak, A., Paliwoda, A., & Błasiak, J. (2019). Anti-proliferative, pro-apoptotic and anti-oxidative activity of Lactobacillus and Bifidobacterium strains: A review of mechanisms and therapeutic perspectives. *Critical Reviews in Food Science and Nutrition*, *59*(21), 3456–3467. DOI: 10.1080/10408398.2018.1494539 PMID: 30010390

Nozari, H., Szmelter-Jarosz, A., & Ghahremani-Nahr, J. (2021). The Ideas of Sustainable and Green Marketing Based on the Internet of Everything—The Case of the Dairy Industry. In *Future Internet* (Vol. 13, Issue 10). DOI: 10.3390/fi13100266

Nyokabi, S. N., de Boer, I. J. M., Luning, P. A., Korir, L., Lindahl, J., Bett, B., & Oosting, S. J. (2021). Milk quality along dairy farming systems and associated value chains in Kenya: An analysis of composition, contamination and adulteration. *Food Control*, *119*, 107482. DOI: 10.1016/j.foodcont.2020.107482

Oerlemans, M. M., Akkerman, R., Ferrari, M., Walvoort, M. T., & de Vos, P. (2021). Benefits of bacteria-derived exopolysaccharides on gastrointestinal microbiota, immunity and health. *Journal of Functional Foods*, *76*, 104289. DOI: 10.1016/j.jff.2020.104289

Ordieres-Meré, J., Gutierrez, M., & Villalba-Díez, J. (2023). Toward the industry 5.0 paradigm: Increasing value creation through the robust integration of humans and machines. *Computers in Industry*, *150*, 103947. DOI: 10.1016/j.compind.2023.103947

Othman, S., Mavani, N. R., Hussain, M. A., Abd Rahman, N., & Ali, J. M. (2023). Artificial intelligence-based techniques for adulteration and defect detections in food and agricultural industry: A review. *Journal of Agriculture and Food Research*, *12*, 100590. DOI: 10.1016/j.jafr.2023.100590

Owusu-Kwarteng, J., Akabanda, F., Agyei, D., & Jespersen, L. (2020). Microbial Safety of Milk Production and Fermented Dairy Products in Africa. In *Microorganisms* (Vol. 8, Issue 5). DOI: 10.3390/microorganisms8050752

Pancholi, N., Gajera, H., & Shah, D. (2023). Improving quality of maintenance task for milk powder manufacturing unit through TOPSIS. *Journal of Quality in Maintenance Engineering*, *29*(1), 27–49. DOI: 10.1108/JQME-04-2021-0028

Pandey, P., Pandey, A., Yan, L., Wang, D., Pandey, V., Meikap, B. C., Huo, J., Zhang, R., & Pandey, P. K. (2021). Dairy waste and potential of small-scale biogas digester for rural energy in India. *Applied Sciences (Basel, Switzerland)*, *11*(22), 10671. DOI: 10.3390/app112210671

Pappas, N., Caputo, A., Pellegrini, M. M., Marzi, G., & Michopoulou, E. (2021). The complexity of decision-making processes and IoT adoption in accommodation SMEs. *Journal of Business Research, 131*, 573–583. DOI: 10.1016/j.jbusres.2021.01.010

Pasupuleti, V., Thuraka, B., Kodete, C. S., & Malisetty, S. (2024). Enhancing Supply Chain Agility and Sustainability through Machine Learning: Optimization Techniques for Logistics and Inventory Management. In *Logistics* (Vol. 8, Issue 3). DOI: 10.3390/logistics8030073

Patel, A. S., Brahmbhatt, M. N., Bariya, A. R., Nayak, J. B., & Singh, V. K. (2023). Blockchain technology in food safety and traceability concern to livestock products. *Heliyon, 9*(6), e16526. DOI: 10.1016/j.heliyon.2023.e16526 PMID: 37265625

Patil, G. B., Wani, S. P., Bafna, P. S., Bagul, V. S., Kalaskar, M. G., & Mutha, R. E. (2024). Milk adulteration: From detection to health impact. *Food and Humanity*, 100339.

Pereira, T. D. S., dos Santos, D. M., Andre, R. D. S., & Correa, D. S. (2024). Zein/MnO2 Nanosheet Composites Integrated with a Smartphone for Colorimetric Sensors for On-Site Detection of Adulterants in Milk. *ACS Applied Nano Materials*.

Peris, M., & Escuder-Gilabert, L. (2016). Electronic noses and tongues to assess food authenticity and adulteration. *Trends in Food Science & Technology, 58*, 40–54. DOI: 10.1016/j.tifs.2016.10.014

Pettitt, D. J. (2020). Xanthan gum. In *Food hydrocolloids* (pp. 127–149). CRC Press.

Poghossian, A., Geissler, H., & Schöning, M. J. (2019). Rapid methods and sensors for milk quality monitoring and spoilage detection. *Biosensors & Bioelectronics, 140*, 111272. DOI: 10.1016/j.bios.2019.04.040 PMID: 31170654

Pourjafar, H., Ansari, F., Sadeghi, A., Samakkhah, S. A., & Jafari, S. M. (2023). Functional and health-promoting properties of probiotics' exopolysaccharides; isolation, characterization, and applications in the food industry. *Critical Reviews in Food Science and Nutrition, 63*(26), 8194–8225. DOI: 10.1080/10408398.2022.2047883 PMID: 35266799

Prasad, S., & Purohit, S. R. (2023). Microbial exopolysaccharide: Sources, stress conditions, properties and application in food and environment: A comprehensive review. *International Journal of Biological Macromolecules, 242*, 124925. DOI: 10.1016/j.ijbiomac.2023.124925 PMID: 37236568

Prete, R., Alam, M. K., Perpetuini, G., Perla, C., Pittia, P., & Corsetti, A. (2021). Lactic acid bacteria exopolysaccharides producers: A sustainable tool for functional foods. *Foods, 10*(7), 1653. DOI: 10.3390/foods10071653 PMID: 34359523

Pu, Y., Pérez-Marín, D., O'Shea, N., & Garrido-Varo, A. (2021). Recent advances in portable and handheld NIR spectrometers and applications in milk, cheese and dairy powders. *Foods*, *10*(10), 2377. DOI: 10.3390/foods10102377 PMID: 34681426

Pu, Y.-Y., O'Donnell, C., Tobin, J. T., & O'Shea, N. (2020). Review of near-infrared spectroscopy as a process analytical technology for real-time product monitoring in dairy processing. *International Dairy Journal*, *103*, 104623. DOI: 10.1016/j.idairyj.2019.104623

Qin, C., Liu, L., Wang, Y., Leng, T., Zhu, M., Gan, B., Xie, J., Yu, Q., & Chen, Y. (2022). Advancement of omics techniques for chemical profile analysis and authentication of milk. *Trends in Food Science & Technology*, *127*, 114–128. DOI: 10.1016/j.tifs.2022.06.001

Rahman, A., Habib, M. R., Ali, M. Y., Islam, M. A., & Rashid, M. H. U. (2017). Physico-chemical analysis and detection of adulteration in raw milk collected from Goalas of different places of sadar upazila in Mymensingh district. *Research in Agriculture Livestock and Fisheries*, *4*(2), 99–106. DOI: 10.3329/ralf.v4i2.33721

Raj, T., Chandrasekhar, K., Kumar, A. N., & Kim, S.-H. (2022). Recent biotechnological trends in lactic acid bacterial fermentation for food processing industries. *Systems Microbiology and Biomanufacturing*, *2*(1), 14–40. DOI: 10.1007/s43393-021-00044-w

Rathakrishnan, T., Suntharalingam, C., Yi Shern, L., Loong, S. K., Lee, H., Tee, K., Thanarajoo, S., & Balasundram, S. (2022). Adoption of Sustainable Agricultural Practices Among Smallholder Dairy Farmers in Malaysia: Contributing Factors and Smart Farming Prospects. *Journal of Smart Science and Technology*, *2*(2), 16–31. DOI: 10.24191/jsst.v2i2.34

Rebechi, S. R., Vélez, M. A., Vaira, S., & Perotti, M. C. (2016). Adulteration of Argentinean milk fats with animal fats: Detection by fatty acids analysis and multivariate regression techniques. *Food Chemistry*, *192*, 1025–1032. DOI: 10.1016/j.foodchem.2015.07.107 PMID: 26304443

Rejeb, A., Keogh, J., Zailani, S., Treiblmaier, H., & Rejeb, K. (2020). Blockchain Technology in the Food Industry: A Review of Potentials, Challenges and Future Research Directions. *Logistics (Basel)*, *4*(4), 27. DOI: 10.3390/logistics4040027

Romulo, A., & Meindrawan, B. (2021). Effect of dairy and non-dairy ingredients on the physical characteristic of ice cream. IOP Conference Series: Earth and Environmental Science. PMID: 34845955

Sahoo, S., & Lo, C.-Y. (2022). Smart manufacturing powered by recent technological advancements: A review. *Journal of Manufacturing Systems*, *64*, 236–250. DOI: 10.1016/j.jmsy.2022.06.008

Salah, B., Alsamhan, A. M., Khan, S., & Ruzayqat, M. (2021). Designing and Developing a Smart Yogurt Filling Machine in the Industry 4.0 Era. *Machines (Basel)*, *9*(11), 300. DOI: 10.3390/machines9110300

Sarkar, D., Ranjan, R., Marwaha, S., Sahoo, A., & Naha, S. (2024). Multiplex PCR for detection of camel milk adulteration with cattle and goat milk. *International Dairy Journal*, *154*, 105922. DOI: 10.1016/j.idairyj.2024.105922

Sar, T., Harirchi, S., Ramezani, M., Bulkan, G., Akbas, M. Y., Pandey, A., & Taherzadeh, M. J. (2022). Potential utilization of dairy industries by-products and wastes through microbial processes: A critical review. *The Science of the Total Environment*, *810*, 152253. DOI: 10.1016/j.scitotenv.2021.152253 PMID: 34902412

Schiller, E., Aidoo, A., Fuhrer, J., Stahl, J., Ziörjen, M., & Stiller, B. (2022). Landscape of IoT security. *Computer Science Review*, *44*, 100467. DOI: 10.1016/j.cosrev.2022.100467

Scudino, H., Guimarães, J. T., Ramos, G. L. P. A., Santillo, A., Pimentel, T. C., Cruz, A. G., & Albenzio, M. (2023). High-intensity ultrasound in cheese processing. *Current Opinion in Food Science*, *50*, 101004. DOI: 10.1016/j.cofs.2023.101004 PMID: 37087267

Sen, S., Dundar, Z., Uncu, O., & Ozen, B. (2021). Potential of Fourier-transform infrared spectroscopy in adulteration detection and quality assessment in buffalo and goat milks. *Microchemical Journal*, *166*, 106207. DOI: 10.1016/j.microc.2021.106207

Shahin, M., Chen, F. F., Hosseinzadeh, A., & Zand, N. (2023). Using machine learning and deep learning algorithms for downtime minimization in manufacturing systems: An early failure detection diagnostic service. *International Journal of Advanced Manufacturing Technology*, *128*(9), 3857–3883. DOI: 10.1007/s00170-023-12020-w

Shamsuddoha, M., & Nasir, T. (2024). Utilizing dairy waste processing for organic agricultural production: A sustainable approach to producing organic goods. *Processes (Basel, Switzerland)*, *12*(11), 2521. DOI: 10.3390/pr12112521

Shao, Y., Yuan, Y., Xi, Y., Zhao, T., & Ai, N. (2023). Effects of Homogenization on Organoleptic Quality and Stability of Pasteurized Milk Samples. *Agriculture*, *13*(1), 205. DOI: 10.3390/agriculture13010205

Sharma, A., Batra, N., Garg, A., & Saxena, A. (2017). Food adulteration: A review. *International Journal for Research in Applied Science and Engineering Technology*, *5*(11), 686–689. DOI: 10.22214/ijraset.2017.3129

Sharma, R., Rajput, Y. S., & Barui, A. K. (2012). *Detection of adulterants in milk: laboratory manual*. NDRI.

Shawky, E., Nahar, L., Nassief, S. M., Sarker, S. D., & Ibrahim, R. S. (2024). Dairy products authentication with biomarkers: A comprehensive critical review. *Trends in Food Science & Technology*, *147*, 104445. DOI: 10.1016/j.tifs.2024.104445

Shingh, S., Kamalvanshi, V., Ghimire, S., & Basyal, S. (2020). Dairy Supply Chain System Based on Blockchain Technology. Asian Journal of Economics, Business and Accounting, 13-19. DOI: 10.9734/ajeba/2020/v14i230189

Singh, A., Chandra, G., Aggarwal, A., & Kumar, P. (2012). Adulteration detection in milk. *Res News U*, *5*, 52–55.

Sinha, B. B., & Dhanalakshmi, R. (2022). Recent advancements and challenges of Internet of Things in smart agriculture: A survey. *Future Generation Computer Systems*, *126*, 169–184. DOI: 10.1016/j.future.2021.08.006

Smith, L., & Wang, Z. (2020). Waste reduction in dairy processing through IoT automation. *Journal of Sustainable Agriculture*, *12*(2), 200–214.

Smith, N. W., Fletcher, A. J., Hill, J. P., & McNabb, W. C. (2022). Modeling the contribution of milk to global nutrition. *Frontiers in Nutrition*, *8*, 716100. DOI: 10.3389/fnut.2021.716100 PMID: 35096919

Soon, J. M., & Abdul Wahab, I. R. (2021). Global food recalls and alerts associated with labelling errors and its contributory factors. *Trends in Food Science & Technology*, *118*, 791–798. DOI: 10.1016/j.tifs.2021.11.001

Soon, J. M., Brazier, A. K. M., & Wallace, C. A. (2020). Determining common contributory factors in food safety incidents – A review of global outbreaks and recalls 2008–2018. *Trends in Food Science & Technology*, *97*, 76–87. DOI: 10.1016/j.tifs.2019.12.030

Sowmya, N., & Ponnusamy, V. (2021). Development of Spectroscopic Sensor System for an IoT Application of Adulteration Identification on Milk Using Machine Learning. *IEEE Access : Practical Innovations, Open Solutions*, *9*, 53979–53995. DOI: 10.1109/ACCESS.2021.3070558

Spink, J. W. (2019). *Food fraud prevention: introduction, implementation, and management*. Springer Nature. DOI: 10.1007/978-1-4939-9621-6

Su, J.-J., Ding, S.-T., & Chung, H.-C. (2020). Establishing a smart farm-scale piggery wastewater treatment system with the internet of things (IoT) applications. *Water (Basel), 12*(6), 1654. DOI: 10.3390/w12061654

Suman, M., Cavanna, D., Sammarco, G., Lambertini, F., & Loffi, C. (2021). Fighting food frauds exploiting chromatography-mass spectrometry technologies: Scenario comparison between solutions in scientific literature and real approaches in place in industrial facilities. *Trends in Analytical Chemistry, 142*, 116305. DOI: 10.1016/j.trac.2021.116305

Surber, G., Schäper, C., Wefers, D., Rohm, H., & Jaros, D. (2021). Exopolysaccharides from Lactococcus lactis affect manufacture, texture and sensory properties of concentrated acid milk gel suspensions (fresh cheese). *International Dairy Journal, 112*, 104854. DOI: 10.1016/j.idairyj.2020.104854

Suryawanshi, N., Naik, S., Eswari, J. S., Nazir, F., Saeed, M., Abbas, A., . . . Nasir, M. (2022). Food science and applied biotechnology. *Foo d Science and Applied Biotechnology, 2022.*

Syamala, A., Dubey, K., & Salunke, P. (2024). Milk and Dairy Products Analogues. In *Food Analogues: Emerging Methods and Challenges* (pp. 35-74). Springer. DOI: 10.1007/978-3-031-69872-9_2

Syrmos, E., Bechtsis, D., Vlachos, D., Papapanagiotakis, G., Todi, T., Papaspyropoulou, M., Georgakidis, K., & Sfitis, N. (2023). In Rabie, K., Kharel, R., Mohjazi, L., Elganimi, T., & Khan, W. U. (Eds.), *An IoT Monitoring System for Dairy Products BT - IoT as a Service* (pp. 12–22). Springer Nature Switzerland.

Taheri, R., Shojafar, M., Alazab, M., & Tafazolli, R. (2021). Fed-IIoT: A Robust Federated Malware Detection Architecture in Industrial IoT. *IEEE Transactions on Industrial Informatics, 17*(12), 8442–8452. DOI: 10.1109/TII.2020.3043458

Talbi, C., Elmarrkechy, S., Youssfi, M., Bouzroud, S., Belfquih, M., Sifou, A., . . . Bouyahya, A. (2023). Bacterial exopolysaccharides: from production to functional features. *Progress In Microbes & Molecular Biology, 6*(1).

Tan, J. (2019). Structuring semisolid foods. *Rheology of Semisolid Foods*, 167-201.

Tan, A., & Ngan, P. T. (2020). A proposed framework model for dairy supply chain traceability. *Sustainable Futures : An Applied Journal of Technology, Environment and Society, 2*, 100034. DOI: 10.1016/j.sftr.2020.100034

Tangorra, F. M., Buoio, E., Calcante, A., Bassi, A., & Costa, A. (2024). Internet of Things (IoT): Sensors Application in Dairy Cattle Farming. *Animals (Basel), 14*(21), 3071. DOI: 10.3390/ani14213071 PMID: 39518794

Tehrani, T., Pont, L., & Benavente, F. (2024). Rapid detection and quantification of milk adulteration using MALDI-MS protein profiling and multivariate calibration. *Journal of Food Composition and Analysis, 130*, 106147. DOI: 10.1016/j.jfca.2024.106147

Theissler, A., Pérez-Velázquez, J., Kettelgerdes, M., & Elger, G. (2021). Predictive maintenance enabled by machine learning: Use cases and challenges in the automotive industry. *Reliability Engineering & System Safety, 215*, 107864. DOI: 10.1016/j.ress.2021.107864

Theodorakopoulos, L., Theodoropoulou, A., & Stamatiou, Y. (2024). A State-of-the-Art Review in Big Data Management Engineering: Real-Life Case Studies, Challenges, and Future Research Directions. In *Eng* (Vol. 5, Issue 3, pp. 1266–1297). DOI: 10.3390/eng5030068

Tian, H., Chen, S., Li, D., Lou, X., Chen, C., & Yu, H. (2022). Simultaneous detection for adulterations of maltodextrin, sodium carbonate, and whey in raw milk using Raman spectroscopy and chemometrics. *Journal of Dairy Science, 105*(9), 7242–7252. DOI: 10.3168/jds.2021-21082 PMID: 35863924

Tian, H., Xiong, J., Chen, S., Yu, H., Chen, C., Huang, J., Yuan, H., & Lou, X. (2023). Rapid identification of adulteration in raw bovine milk with soymilk by electronic nose and headspace-gas chromatography ion-mobility spectrometry. *Food Chemistry: X, 18*, 100696. DOI: 10.1016/j.fochx.2023.100696 PMID: 37187488

Titova, T., & Nachev, V. (2020). Electronic tongue" in the food industry. *Food Science and Applied Biotechnology, 3*(1), 71–76. DOI: 10.30721/fsab2020.v3.i1.74

Tiwari, S., Kavitake, D., Devi, P. B., & Shetty, P. H. (2021). Bacterial exopolysaccharides for improvement of technological, functional and rheological properties of yoghurt. *International Journal of Biological Macromolecules, 183*, 1585–1595. DOI: 10.1016/j.ijbiomac.2021.05.140 PMID: 34044028

Trakadas, P., Simoens, P., Gkonis, P., Sarakis, L., Angelopoulos, A., Ramallo-González, A. P., Skarmeta, A., Trochoutsos, C., Calvo, D., Pariente, T., Chintamani, K., Fernandez, I., Irigaray, A. A., Parreira, J. X., Petrali, P., Leligou, N., & Karkazis, P. (2020). An Artificial Intelligence-Based Collaboration Approach in Industrial IoT Manufacturing: Key Concepts, Architectural Extensions and Potential Applications. *Sensors (Basel), 20*(19), 5480. Advance online publication. DOI: 10.3390/s20195480 PMID: 32987911

Tribby, D., & Teter, V. (2023). Yogurt. In *The sensory evaluation of dairy products* (pp. 199–234). Springer. DOI: 10.1007/978-3-031-30019-6_8

Tsenkova, R., & Muncan, J. (2022). Milk quality management: state of the art and contemporary requirements. In *Aquaphotomics for Bio-diagnostics in Dairy*. Springer. DOI: 10.1007/978-981-16-7114-2_1

Uzoigwe, D. (2024). Evaluating the Effectiveness of Reliability-Centered Maintenance Programs in Food and Beverage Manufacturing Facilities; A Review. *International Journal of Research and Innovation in Applied Science*, *9*(2), 204–232. Advance online publication. DOI: 10.51584/IJRIAS.2024.90218

Valand, R., Tanna, S., Lawson, G., & Bengtström, L. (2020). A review of Fourier Transform Infrared (FTIR) spectroscopy used in food adulteration and authenticity investigations. *Food Additives & Contaminants. Part A, Chemistry, Analysis, Control, Exposure & Risk Assessment*, *37*(1), 19–38. DOI: 10.1080/19440049.2019.1675909 PMID: 31613710

Varma, V. S., Parajuli, R., Scott, E., Canter, T., Lim, T. T., Popp, J., & Thoma, G. (2021). Dairy and swine manure management – Challenges and perspectives for sustainable treatment technology. *The Science of the Total Environment*, *778*, 146319. DOI: 10.1016/j.scitotenv.2021.146319 PMID: 33721638

Vasconcelos, H., Matias, A., Jorge, P., Saraiva, C., Mendes, J., Araújo, J., & Coelho, L. C. (2021). Optical biosensor for the detection of hydrogen peroxide in milk. *Chemistry Proceedings*, *5*(1), 55.

Vejan, P., Khadiran, T., Abdullah, R., & Ahmad, N. (2021). Controlled release fertilizer: A review on developments, applications and potential in agriculture. *Journal of Controlled Release*, *339*, 321–334. DOI: 10.1016/j.jconrel.2021.10.003 PMID: 34626724

Venkatesh, V. G., Kang, K., Wang, B., Zhong, R. Y., & Zhang, A. (2020). System architecture for blockchain based transparency of supply chain social sustainability. *Robotics and Computer-integrated Manufacturing*, *63*, 101896. DOI: 10.1016/j.rcim.2019.101896

Vern, P., Panghal, A., Mor, R. S., & Kamble, S. S. (2024). Blockchain technology in the agri-food supply chain: a systematic literature review of opportunities and challenges. *Management Review Quarterly*, 1-33.

Vinciguerra, L. L., Marcelo, M. C., Motta, T. M., Meneghini, L. Z., Bergold, A. M., & Ferrão, M. F. (2019). Chemometric tools and FTIR-ATR spectroscopy applied in milk adulterated with cheese whey. *Quimica Nova*, *42*(3), 249–254. DOI: 10.21577/0100-4042.20170337

Vinothkanna, A., Dar, O. I., Liu, Z., & Jia, A. Q. (2024). Advanced detection tools in food fraud: A systematic review for holistic and rational detection method based on research and patents. *Food Chemistry*, *446*, 138893. DOI: 10.1016/j.foodchem.2024.138893 PMID: 38432137

Visciano, P., & Schirone, M. (2021). Food frauds: Global incidents and misleading situations. *Trends in Food Science & Technology*, *114*, 424–442. DOI: 10.1016/j.tifs.2021.06.010

Wangen, S. R., Zhang, F., Fadul-Pacheco, L., da Silva, T. E., & Cabrera, V. E. (2021). Improving farm decisions: The application of data engineering techniques to manage data streams from contemporary dairy operations. *Livestock Science*, *250*, 104602. DOI: 10.1016/j.livsci.2021.104602

Wang, L., Gu, Y., & Lv, Z. (2022). Processing properties of yogurt as affected by the EPS produced by Leuconostoc mesenteroides XR1. *International Journal of Food Science & Technology*, *57*(7), 4076–4085. DOI: 10.1111/ijfs.15722

Wang, L., Tian, H., Liu, W., Zheng, H., Wu, H., Guan, Y., Lu, Q., & Lv, Z. (2023). Effects of EPS-producing Leuconostoc mesenteroides XR1 on texture, rheological properties, microstructure and volatile flavor of fermented milk. *Food Bioscience*, *56*, 103371. DOI: 10.1016/j.fbio.2023.103371

Werning, M. L., Hernández-Alcántara, A. M., Ruiz, M. J., Soto, L. P., Dueñas, M. T., López, P., & Frizzo, L. S. (2022). Biological functions of exopolysaccharides from lactic acid bacteria and their potential benefits for humans and farmed animals. *Foods*, *11*(9), 1284. DOI: 10.3390/foods11091284 PMID: 35564008

Whyte, S. T., Omoyiola, B. O., & Okoni, B. (2022). Use of Blockchain Technology in Data Integrity Assurance. *Available at SSRN* 4043164. DOI: 10.2139/ssrn.4043164

Windarsih, A., Arifah, M. F., Utami, I. D., Suratno, , & Rohman, A. (2024). Detection of goat milk adulteration in horse milk using LC-HRMS-based non-targeted metabolomics and chemometrics. *Chemicke Zvesti*, *78*(2), 809–821. DOI: 10.1007/s11696-023-03123-5

Wu, J.-Y., & Hsiao, H.-I. (2021). Food quality and safety risk diagnosis in the food cold chain through failure mode and effect analysis. *Food Control*, *120*, 107501. DOI: 10.1016/j.foodcont.2020.107501

Xu, J., Guo, S., Xie, D., & Yan, Y. (2020). Blockchain : A new safeguard for agri-foods. *Artificial Intelligence in Agriculture*, *4*, 153–161. DOI: 10.1016/j.aiia.2020.08.002

Xu, Q., Wang, H., Ren, Y., Sun, M., Zhang, T., Li, H., & Liu, X. (2024). Functionality and application of emulsion gels in fat replacement strategies for dairy products. *Trends in Food Science & Technology, 152*, 104673. DOI: 10.1016/j.tifs.2024.104673

Xu, Z., Guo, Q., Zhang, H., Xiong, Z., Zhang, X., & Ai, L. (2021). Structural characterisation of EPS of Streptococcus thermophilus S-3 and its application in milk fermentation. *International Journal of Biological Macromolecules, 178*, 263–269. DOI: 10.1016/j.ijbiomac.2021.02.173 PMID: 33639187

Yaman, H. (2020, August). A rapid method for detection adulteration in goat milk by using vibrational spectroscopy in combination with chemometric method s. *Journal of Food Science and Technology, 57*(8), 3091–3098. DOI: 10.1007/s13197-020-04342-4 PMID: 32624611

Yang, J., Zheng, N., Soyeurt, H., Yang, Y., & Wang, J. (2019). Detection of plant protein in adulterated milk using nontargeted nano-high-performance liquid chromatography–tandem mass spectroscopy combined with principal component analysis. *Food Science & Nutrition, 7*(1), 56–64. DOI: 10.1002/fsn3.791 PMID: 30680159

Yang, L., Wang, L., Chi, Y., & Chi, Y. (2024). Effect of Whole Egg Liquid on Physicochemical, Quality, Fermentation and Sensory Characteristics of Yogurt. *Foods, 13*(2), 321. DOI: 10.3390/foods13020321 PMID: 38275687

Yang, S., Xu, X., Peng, Q., Ma, L., Qiao, Y., & Shi, B. (2023). Exopolysaccharides from lactic acid bacteria, as an alternative to antibiotics, on regulation of intestinal health and the immune system. *Animal Nutrition, 13*, 78–89. DOI: 10.1016/j.aninu.2023.02.004 PMID: 37025257

Yang, X., Li, A., Li, X., Sun, L., & Guo, Y. (2020). An overview of classifications, properties of food polysaccharides and their links to applications in improving food textures. *Trends in Food Science & Technology, 102*, 1–15. DOI: 10.1016/j.tifs.2020.05.020

Yang, X., Ren, Y., & Li, L. (2022). The relationship between charge intensity and bioactivities/processing characteristics of exopolysaccharides from lactic acid bacteria. *Lebensmittel-Wissenschaft + Technologie, 153*, 112345. DOI: 10.1016/j.lwt.2021.112345

Yashaswini, N., Suraj, L., Nayak, S., Khosla, A., & Manjunatha, C. (2022). Construction, Working, and Applications of E-Tongue: A Versatile Tool for All Tastes? *ECS Transactions, 107*(1), 20193–20211. DOI: 10.1149/10701.20193ecst

Yildiz, H., & Karatas, N. (2018). Microbial exopolysaccharides: Resources and bioactive properties. *Process Biochemistry (Barking, London, England)*, *72*, 41–46. DOI: 10.1016/j.procbio.2018.06.009

Yin, X., Li, J., Zhu, L., & Zhang, H. (2024). Advances in the formation mechanism of set-type plant-based yogurt gel: A review. *Critical Reviews in Food Science and Nutrition*, *64*(26), 9412–9431. PMID: 37203992

Yin, Z.-y., Yuan, Y., Zhang, R., Gan, J., Yu, L., Qiu, X., Chen, R., & Wang, Q. (2024). Understanding Bacillus response to salt stress: Growth inhibition, enhanced EPS secretion, and molecular adaptation mechanisms. *Process Biochemistry (Barking, London, England)*, *146*, 412–422. DOI: 10.1016/j.procbio.2024.09.023

Yue, Y., Wang, Y., Yang, R., Liu, D., Cheng, Y., Li, S., Sun, L., Li, B., & Huo, G. (2023). Exopolysaccharides of Bifidobacterium longum subsp. infantis E4 on the immune and anti-inflammatory effects in vitro. *Journal of Functional Foods*, *107*, 105699. DOI: 10.1016/j.jff.2023.105699

Zadeike, D., & Degutyte, R. (2023). Recent advances in acoustic technology in food processing. *Foods*, *12*(18), 3365. DOI: 10.3390/foods12183365 PMID: 37761074

Zarin, I., Talukder, A. K., Hoque, M. N., Rahman, M., Haider, G., Rahman, A. N. M. A., & Das, Z. C. (2023). Current scenario on reproductive management and performances of dairy cows in Manikganj district. *Annals of Bangladesh Agriculture*, *27*(2), 129–144. DOI: 10.3329/aba.v27i2.72541

Żbik, K., Onopiuk, A., Górska-Horczyczak, E., & Wierzbicka, A. (2024). Trends and Opportunities in the Dairy Industry: A2 Milk and Processing Methods. *Applied Sciences (Basel, Switzerland)*, *14*(15), 6513. DOI: 10.3390/app14156513

Zhang, K., Tang, H., Farid, M. S., Xiang, F., & Li, B. (2024). Effect of Lactobacillus helveticus exopolysaccharides molecular weight on yogurt gel properties and its internal mechanism. *International Journal of Biological Macromolecules*, *262*, 130006. DOI: 10.1016/j.ijbiomac.2024.130006 PMID: 38331067

Zhang, K., Zhang, T.-T., Guo, R.-R., Ye, Q., Zhao, H.-L., & Huang, X.-H. (2023). The regulation of key flavor of traditional fermented food by microbial metabolism: A review. *Food Chemistry: X*, *19*, 100871. DOI: 10.1016/j.fochx.2023.100871 PMID: 37780239

Zhang, T., Liu, W., Lu, H., Cheng, T., Wang, L., Wang, G., & Chen, W. (2023). Lactic acid bacteria in relieving constipation: Mechanism, clinical application, challenge, and opportunity. *Critical Reviews in Food Science and Nutrition*, 1–24. PMID: 37971876

Zhao, V. X. T., Wong, T. I., Zheng, X. T., Tan, Y. N., & Zhou, X. (2020). Colorimetric biosensors for point-of-care virus detections. *Materials Science for Energy Technologies*, *3*, 237–249. DOI: 10.1016/j.mset.2019.10.002 PMID: 33604529

Zhou, Y., Cui, Y., & Qu, X. (2019). Exopolysaccharides of lactic acid bacteria: Structure, bioactivity and associations: A review. *Carbohydrate Polymers*, *207*, 317–332. DOI: 10.1016/j.carbpol.2018.11.093 PMID: 30600013

Zhu, C., Fu, X., Zhang, J., Qin, K., & Wu, C. (2022). Review of portable near infrared spectrometers: Current status and new techniques. *Journal of Near Infrared Spectroscopy*, *30*(2), 51–66. DOI: 10.1177/09670335211030617

Ziyaina, M., Rasco, B., & Sablani, S. S. (2020). Rapid methods of microbial detection in dairy products. *Food Control*, *110*, 107008. DOI: 10.1016/j.foodcont.2019.107008

About the Contributors

Poliana Mendes de Souza is a professor in the Animal Food Industry and Valorization, at the University Mohammed VI Polytechnic (UM6P). She holds a PhD in Food Technology from UPV (Spain, 2012), which included a research period at MRI (Germany), a Master in Food Engineering from UNICAMP (Brazil, 2007), and a Bachelor in Food Engineering from PUCPR (Brazil, 2006). Dr. Souza served as an associate professor at the Science and Technology Institute of the Federal University of Jequitinhonha and Mucuri Valley (UFVJM), Brazil, from 2013 to 2024. During her tenure, she led the Nutrition Sciences Graduate Program, and the Emerging Food Processing Technologies Research Group, and was the founder and first coordinator of the Master Program in Food Science and Technology. Her teaching expertise includes food science and technology, with a focus on project-based learning and learning by doing methodologies. Since 2002, Dr. Souza's research has focused on emerging technologies for the conservation of food matrices. Her recent research projects aimed to reduce microbiological contamination in meat and dairy products while surveying their sensorial and nutritional quality. Key projects included quantitative and predictive microbiology of meat and milk, the study of irradiation technologies (UV-C and gamma irradiation), and the valorization of byproducts from the food industry. Her ongoing projects at ASARI will deal with camel milk valorization and conservation, mainly the development and the quality assessment of new camel milk products. Dr. Souza has authored papers, book chapters, and conference publications. She advised graduate and undergraduate students and collaborated with local industries and community partners, mainly in Brazil, Uruguay, Mexico, and Spain.

***

Nagamalli Arasavalli is a distinguished academic and visionary researcher, serving as Assistant Professor in the Department of Electronics and Communication

Engineering at Koneru Lakshmaiah Education Foundation (Deemed to be University), Guntur, Andhra Pradesh, India. With a Ph.D. from Andhra University (2023) and an M.Tech in VLSI Systems Design from JNTU, Kakinada, her scholarly foundation is both robust and exemplary. A trailblazer in technological innovation, Dr. Arasavalli is a Medium-Term Research Project Member with INUP-CeNSE, IISc Bangalore. Her groundbreaking contributions, underscored by patented research advancements, have earned her prestigious accolades, including the "CHANAKYA" Award (ICERT, USA & India, 2023) and the "Dr. Sarvepalli Radhakrishnan Award" (2024) for excellence in technology and societal impact. She further elevates her professional stature as a reviewer for esteemed journals such as IEEE ACCESS and JETIR (UGC-approved). Her domain expertise spans cutting-edge areas, including MEMS gas sensor fabrication, nanomaterials synthesis for biosensors, advanced micro and nano-fabrication techniques, IoT sensor design, VLSI architectures for cryptography, and the application of machine learning and deep learning in disease diagnostics. Driven by an unwavering commitment to innovation and societal advancement, Dr. Arasavalli continues to shape the future of technology through impactful research and academic leadership.

Leena Arya is working as a Professor at the Department of Computer Science and Engineering at Koneru Lakshmaiah Education Foundation Deemed to be University, Vaddeswaram, 22nd NIRF Ranking University, Guntur, AP, India. She completed her Ph.D. from IIT Roorkee in 2012. She has published over 55 research papers in Scopus and SCI-indexed international journals. She handled a research project for the Department of Science and Technology, DST, New Delhi, as Principal Investigator under the Women Scientists Scheme (WOS-A). She has been granted a UK design and Australian Government Innovation patents. She has been a reviewer of many prestigious international journals and conferences. Her research area includes wireless communication, the Internet of Things (IoT), network security, and artificial intelligence.

Latha Banda is working as a Professor & Head CSE, in JEMTEC, Greater Noida, Uttar Pradesh. She did Ph.D from Jawaharlal Nehru University, New Delhi. She has more than 19 years of experience in teaching and research. She has published more than 40 (research papers in) journals and International conferences, submitted 2 projects in DeiTY and DST, and 4 patents. She has considerable experience in administration, organized more than 80+ seminars, workshops, Industrial visits and many more. She organized 5 International conferences and 2 National conferences. Her research interests are Recommender Systems, Machine Learning and Artificial Intelligence. She is the reviewer of many Springer and ACM journals and PC member of various International conferences. She has guided 2 PhD and 12 M.Tech

Students, the editor of a Book series on Machine Learning in Security. She has collaborations with Jawaharlal Nehru University and the National University of Malaysia.

Loubna El Mansouri is Professor at the College of Geomatic Sciences and Surveying Engineering at Institute of Agronomy and Veterinary Medicine (IAV) Hassan II, Rabat, Morocco, where she has been since 2004. From 2014 to 2017 she served as Head of Department of Geodetic Sciences at IAV Hassan 2. During 2009-2013 she was a Ph.D. student at "SURAFCE" Laboratory at Science College of LIEGE University (ULG) in Belgium. From 2010 to 2012 she was a visiting staff at "MONTEFIORE" Institute of Computer Science at ULG. She received an engineering degree in Geomatics and Geodetic Sciences from IAV Hassan 2 in 2004 and enrolled as licensed Surveyor. Holds a Doctorate of Sciences in Remote Sensing from the ULG in 2013. Her research interests span both digital image process and remote sensing. Much of her work has been on improving the application of maching learning technique in land use/ land cover data producing for territory management, more over in agriculture and forest domain. Professor Elmansouri published more than 30 papers, coordinator of many R&D projects, and Co-Advisor of Doctorate Studies.

Ahmed Hamad is a renowned food science and nutrition expert with years of experience in teaching, research, and industry. Dr. Ahmed has published numerous papers and books on food hygiene, nutrition, and bioscience and has worked with various government and non-government organizations on food safety and nutrition projects. Ahmed has been teaching and researching food science, Safety, technology, and quality for Twelve (12) years. Ahmed teaches Food Chemistry, Microbiology, Food Technology and processing, Food Safety, and TQM courses in the food industry. Ahmed has many publications (+25) in highly prestigious peer-reviewed journals. He also contributes to many funded scientific projects in food safety, antimicrobial resistance, biofilm formation and control, Meat quality improvement, and environmental pollution. Ahmed participated as a peer reviewer for many prestigious food journals. Ahmed Has a role as a Book Editor for IGI Global and CRC. Ahmed is a member of many professional associations and societies, such as RSPH, IFST, IAFP, AMSA, and ASM. He has also worked as an associate food safety and quality consultant from 2012 till now.

Shashi Mehrotra is a Professor at the Department of Computer Science and Engineering, SRM Institute of Science and Technology, Delhi NCR Campus, Modinagar, Ghaziabad, India. She holds a Ph.D. in Technology from the Birla Institute of Technology, Mesra, an M.Tech in Computer Engineering from ITM

Gurgaon, and an MCA from Madurai Kamaraj University (MKU). With over 25 years of professional experience, Dr. Shashi Mehrotra is deeply engaged in teaching and research in Computer Science. Dr. Shashi Mehrotra has authored numerous research papers published in esteemed international and national journals and in national and international conferences. She has presented her work at global forums, including in countries such as Singapore, Thailand, and Taiwan. Beyond her research, she has actively contributed to the academic community by organising professional events such as special sessions, faculty development programs (FDPs), workshops, and expert lectures. She has delivered expert talks, invited speakers, keynote speaker, and session chair at national and international conferences. Her research interests primarily include computational intelligence, healthcare, and information retrieval.

Ravi Rastogi received his Ph.D. in Computer Science and Engineering in December 2011 from Uttarakhand Technical University, Dehradun, Uttarakhand, India. MS in Computer Science degree in September 2005 from Fairleigh Dickinson University, Teaneck, New Jersey, USA. He is working as a Professor in the Department of Computer Science and Engineering (Honors) at Koneru Lakshmaiah Education Foundation Deemed to be a university. He was a database developer for Microsoft, Seattle, USA, and ING, Philadelphia, USA. His research interests are interconnection networks, stable matching problems, data mining and intelligence in systems, the Internet of Things, and cloud computing. Refereed International Journal publications are 22 and Refereed International Conference Publications are 8.

SatyanarayanaSanku obtained Bachelor of Technology (ECE) from Jawaharlal Technological University, Hyderabad, India. Master of Engineering (Digital Systems) and PhD (ECE) in Wireless Multimedia Sensor Networks from Osmania University, Hyderabad India. Presently working as faculty of ECE Department in Sreenidhi institute of Science and Technology, Hyderabad, India. Research interest includes Internet of Things, Wearable technology, Wireless multimedia Sensor Networks, Machine Learning, Deep learning, etc.

Jeremy Issa Victor is a seasoned project management and consulting expert with over a decade of experience in the camel industry. He provides consultancy services to both small and large camelid operations worldwide, advocating for streamlined export processes and sustainable solutions that harmonize modern methodologies with traditional practices. Victor is a member of the TrustLabel Experts Committee at CamelTrust, contributing his expertise to promote ethical and sustainable practices within the camel sector.

Index

A

B

C

D

E

F

G

I

L

M

O

P

Q

R

S

T

V

Y

www.ingramcontent.com/pod-product-compliance
Lightning Source LLC
LaVergne TN
LVHW081328300926
849161LV00008B/1917
* 9 7 9 8 3 6 9 3 9 9 4 9 1 *